TEACHING SECONDARY GEOGRAPHY

Geography is not only the study of the surface of the planet and the exploration of spatial and human–environment relationships, but also a way of thinking about the world. Guided by the Australian Curriculum and the Professional Standards for Teaching School Geography (GEOGstandards), *Teaching Secondary Geography* provides a comprehensive introduction to both the theory and practice of teaching geography.

This text examines the core geography concepts of place, space, environment, interconnection, scale, change and sustainability. It covers fundamental geographical knowledge and skills, such as working with data, graphicacy, fieldwork and spatial technology, and provides practical guidance on teaching them in the classroom. Each chapter features short-answer and 'Pause and think' questions to enhance understanding of key concepts, and 'Bringing it together' review questions to consolidate learning. Classroom scenarios and a range of information boxes are provided throughout to connect students to additional material.

Written by an author team with extensive teaching experience, *Teaching Secondary Geography* is an exemplary resource for pre-service teachers.

Malcolm McInerney is a lecturer in humanities education at the University of South Australia and the Manager for Humanities Projects in the South Australian Department for Education.

John Butler OAM teaches methodology and curriculum to future teachers of geography at Flinders University.

Susan Caldis is a lecturer in secondary social science and convenor of geography methodology units in the Macquarie School of Education at Macquarie University.

Stephen Cranby (recently retired) taught geography in high schools for 30 years followed by six years teaching geography method at Monash University.

Susanne Jones is a casual academic tutoring in Humanities and Social Sciences in the School of Education at the University of South Australia.

Mick Law is a geography teacher with a passion for embedding geospatial and other digital tools into his teaching and learning.

Alaric Maude is a retired associate professor of geography, affiliated with Flinders University in Adelaide.

Rebecca Nicholas is a deputy principal at a public high school in Queensland, and has taught geography for the last 21 years at a number of schools in both Queensland and Victoria.

TEACHING SECONDARY GEOGRAPHY

Malcolm McInerney, John Butler,
Susan Caldis, Stephen Cranby,
Susanne Jones, Mick Law and
Alaric Maude, Rebecca Nicholas

CAMBRIDGE
UNIVERSITY PRESS

CAMBRIDGE
UNIVERSITY PRESS

University Printing House, Cambridge CB2 8BS, United Kingdom

One Liberty Plaza, 20th Floor, New York, NY 10006, USA

477 Williamstown Road, Port Melbourne, VIC 3207, Australia

314–321, 3rd Floor, Plot 3, Splendor Forum, Jasola District Centre, New Delhi – 110025, India

103 Penang Road, #05–06/07, Visioncrest Commercial, Singapore 238467

Cambridge University Press is part of the University of Cambridge.

It furthers the University's mission by disseminating knowledge in the pursuit of education, learning and research at the highest international levels of excellence.

www.cambridge.org
Information on this title: www.cambridge.org/9781108984638

First published 2022

Cover designed by Cate Furey
Text designed by Anne-Marie Reeves
Typeset by Straive

A catalogue record for this publication is available from the British Library

A catalogue record for this book is available from the National Library of Australia

ISBN 978-1-108-98463-8 Paperback

Additional resources for this publication at www.cambridge.org/highereducation/isbn/9781108984638/resources.

CONTENTS

About the authors   *page* viii
Acknowledgements   x
Guide to online resources   xii

Part 1 Introducing geography   1

1 What is geography?   3
Alaric Maude

   Introduction   3
   What is geography?   3
   The aims of the Australian Curriculum: Geography   4
   Geography's concepts   6
   Conclusion   31
   References   31

2 Core knowledge through case studies   34
John Butler

   Introduction   34
   The Australian Curriculum: Geography   35
   Core knowledge of students at the end of Year 6   36
   Unpacking the core knowledge and suitable case studies for Years 7 to 10   38
   Unpacking the core knowledge and suitable case studies for Years 11 and 12   91
   Conclusion   99
   References   100

Part 2 Geographical skills   103

3 The graphicacy of geography   105
Rebecca Nicholas

   Introduction   105
   What is graphicacy?   106
   Types of graphicacy   113
   Strategies to include graphicacy in the geography classroom   122
   Conclusion   126
   References   126

4 Working with data   129
Rebecca Nicholas

   Introduction   129
   Data in geography   130
   Data in the geography classroom   131
   Data visualisations and infographics   134
   Engaging students with geographic data in the classroom   138
   Conclusion   140
   References   141

5 Fieldwork skills   142
Stephen Cranby

    Introduction   142
    Fieldwork skills in students' wider learning   143
    Fieldwork skills for thinking geographically   147
    Geographic fieldwork skills   152
    Geographic fieldwork tools   157
    Conclusion   160
    References   161

6 Using spatial technology   163
Mick Law

    Introduction   163
    What are geospatial technologies?   164
    The benefits of using geospatial technologies in the classroom   170
    Strategies to effectively bring geospatial technologies into your classroom 172
    Conclusion   181
    References   182

Part 3 Teaching geography   185

7 The inquiry process in geography   187
John Butler and Susan Caldis

    Introduction   187
    Inquiry as part of the Australian Curriculum   187
    Why is an inquiry approach used in teaching and learning geography?   190
    Inquiry and thinking   194
    Using questions   196
    Conclusion   199
    References   200

8 What makes my geography lesson distinctive and powerful?   201
Susan Caldis

    Introduction   201
    The 'what?' and 'why?' of a distinctive and powerful geographical geography lesson 202
    The 'how?' of a distinctive and powerful geographical geography lesson   210
    Conclusion   213
    References   215

9 Fieldwork   218
Stephen Cranby

    Introduction   218
    Centrality of fieldwork   219
    Pedagogy of fieldwork   221
    Practice of fieldwork   229
    Conclusion   247
    References   248

10 The general capabilities' synergy with geography   249
Malcolm McInerney

    Introduction   249
    The general capabilities   251
    Geography and the general capabilities   256
    Conclusion   264
    References   265

11 The importance of planning in geography   267
Malcolm McInerney, Susanne Jones and Susan Caldis

    Introduction   267
    The planning approach   267
    Assessment   273
    Unit planning   282
    Conclusion   282
    References   284

12 The professionally engaged geography teacher   285
Susan Caldis and Alaric Maude

    Introduction   285
    What is professional engagement?   286
    Professional engagement for teachers of geography   286
    Using the GEOGstandards to enact professional engagement in the teaching
    of geography   289
    Using geographical resources to professionally engage with geography
    in the classroom   296
    Professionally engaging with geography beyond the classroom   297
    Conclusion   298
    References   299

Index   301

ABOUT THE AUTHORS

Malcolm McInerney is a teaching academic in humanities education at the University of South Australia and the manager for humanities projects in the South Australian Department for Education. He has taught geography in the South Australian education system since 1976. Malcolm has been actively involved in the promotion of geography in his work as chair of the Australian Geography Teachers Association (2008–13), as a member of the Australian Curriculum: Geography ACARA Advisory Panel (2009–13) and as executive director of Education Services Australia's GeogSpace project. During these years, Malcolm was also greatly involved in the use of spatial technology in geographical education.

John Butler OAM teaches methodology and curriculum to future teachers of geography at Flinders University, having retired after 40 years spent teaching secondary school students in both government and private schools. He has been a teacher and enthusiast of geography all his life, and has been an adviser, a state-wide consultant, an author, an examiner, a curriculum developer and a school administrator. John has received numerous awards, including an Order of Australia Medal (OAM) for his services to education. He has written and contributed to 40 books, mostly for secondary geography. In recent years, he has written for both primary and secondary teachers on websites and in book series.

Susan Caldis recently completed her PhD in geography education and is now a lecturer and convenor of geography methodology units in the Macquarie School of Education at Macquarie University, Sydney. Susan has previously held roles in school-based leadership and national curriculum development for geography. During her doctoral candidature, Susan received a Faculty Award for 'Excellence in Higher Degree Research' and was hosted in Singapore by the Academy of Singapore Teachers as the Outstanding Educator in Residence for Geographical Education. Susan's research focus is on the transformation of practice among pre-service geography teachers as they transition into the profession. Susan is a long-term advocate for geography education, having twice held leadership roles in state and national professional associations. She is also a recently appointed STEM Ambassador for geographical education.

Stephen Cranby (recently retired) taught geography in high schools for 30 years, followed by six years teaching geography method at Monash University. He is a life member of the Geography Teachers' Association of Victoria and a former chairperson of the Australian Geography Teachers Association. He was chief examiner for VCE Geography and has contributed to VCE Geography course design, assessment and advice to teachers over many years. He has contributed to the writing of numerous geography textbooks and professional development workshops at both state and national levels. He has also organised and led more than 100 day and overnight fieldwork exercises with secondary students.

Susanne Jones is a casual academic tutoring in Humanities and Social Sciences in the School of Education at the University of South Australia. She has been a primary and secondary teacher and leader in country and metropolitan schools. Susanne has worked with teachers and leaders during the implementation of the Australian Curriculum, and more recently to enable curriculum change and improve student learning, supporting their work in curriculum, assessment, moderation and writing improvement.

Mick Law is a geography teacher with a passion for embedding geospatial and other digital tools into his teaching and learning. He has been a writer, reviewer and 'critical friend' during the development of several state and national curricula, as well as the author of several textbooks in the field of geography education.

Alaric Maude is a retired associate professor of geography, affiliated with Flinders University in Adelaide. He was the lead writer and then writing coach for the Australian Curriculum: Geography from 2009 to 2013. From that experience, he wrote *Understanding and Teaching the Australian Geography Curriculum for Primary Schools*, to help primary school teachers with the new curriculum. He has also published articles on sustainability and powerful geographical knowledge, and on adapting geography's concepts for use in schools. He was chair of the Academy of Science's National Committee for Geographical Sciences from 2013 to 2017, and led the production of a report titled *Geography: Shaping Australia's Future*.

Rebecca Nicholas is a deputy principal at a public high school in Queensland, and has taught geography for the last 21 years at a number of schools in Queensland and Victoria. She has been involved with geography teaching associations at both the state and national levels, volunteering on the Australian Geography Teachers Association executive for five years. Rebecca has also been involved in the writing and review of the Australian Curriculum: Geography, the 2019 Queensland Senior Geography Syllabus and revisions of the VCAA Senior Geography Syllabus. She has a passion for the use of spatial technologies and digital pedagogy in the classroom, and has presented at a variety of state and national conferences over the last 15 years. Rebecca has been involved in the development of resources to support teachers in both primary and secondary classrooms.

ACKNOWLEDGEMENTS

Mick: For my Mum and Dad who gave me everything.

Alaric: I would like to acknowledge Malcolm McInerney, for sound advice.

John: To my wife Sue, for her constant support and encouragement.

The authors and Cambridge University Press would like to thank the following for permission to reproduce material in this book.

Figure 1.1: © Getty Images/RAW.Exposed; 1.2, 1.5: © Alaric Maude; 1.3, 3.4: Australian Bureau of Meteorology. Licensed under Creative Commons Attribution 3.0 Australia (CC BY 3.0 AU, https://creativecommons.org/licenses/by/3.0/au/); 1.6: Adapted from G. Walker, M. Gilfedder & J. Williams (1999) *Effectiveness of current farming systems in the control of dryland salinity*, Canberra: CSIRO; 2.2: © Google Maps; 2.3: Sourced from Land Information New Zealand. Crown Copyright reserved; 3.2, 7.1, 10.1–10.7 and extracts from the Australian Curriculum: Geography: © Australian Curriculum, Assessment and Reporting Authority (ACARA) 2009 to present, unless otherwise indicated. This material was downloaded from the ACARA website (www.acara.edu.au) (https://www.australiancurriculum.edu.au) (accessed September 2020) and was not modified. The material is licensed under CC BY 4.0 (https://creativecommons.org/licenses/by/4.0/). ACARA does not endorse any product that uses ACARA material or make any representations as to the quality of such products. Any product that uses material published on this website should not be taken to be affiliated with ACARA or have the sponsorship or approval of ACARA. It is up to each person to make their own assessment of the product; 3.3: Australian Bureau of Statistics. Licensed under Creative Commons Attribution 4.0 International licence (https://creativecommons.org/licenses/by/4.0/); 4.1: © Getty Images/Anadolu Agency/Contributor; 6.1: Wikimedia Commons, https://en.wikipedia.org/wiki/File:Snow-cholera-map-1.jpg; 6.2: Reproduced by permission of the publisher, © 2012 by tpack.org; 6.3: Wikimedia Commons, Lefflerd, https://commons.wikimedia.org/wiki/File:The_SAMR_Model.jpg, licensed under CC BY 4.0, https://creativecommons.org/licenses/by/4.0/deed.en.

Extracts from National Committee for Geographical Sciences (2018) *Geography: Shaping Australia's future*. Canberra: Australian Academy of Science: Reproduced with permission.

Extract from International Geographical Union Commission on Geographical Education. (2015). *International Research in Geographical Education*: Reproduced with permission

Extract from Australian Institute for Teaching and School Leadership (AITSL) (2017). Australian Professional Standards for Teachers: © 2018 Education Services Australia Limited as the legal entity for the COAG Education Council (Education Council). Cambridge University Press has reproduced extracts of the Australian

Institute for Teaching and School Leadership's Australian Professional Standards for Teachers in this publication with permission from the copyright owner.

Every effort has been made to trace and acknowledge copyright. The publisher apologises for any accidental infringement and welcomes information that would redress this situation.

GUIDE TO ONLINE RESOURCES

The online student resources for *Teaching Secondary* Geography are available at www
.cambridge.org/highereducation/isbn/9781108984638/resources. Here you can explore
additional activities, weblinks, worksheets, templates, further resources and a glossary
of key terms for each chapter.

 **TEACHING SECONDARY GEOGRAPHY**

Chapter 9 Fieldwork

Glossary

Cooperative learning: in fieldwork, occurs where students work together on shared tasks
with a common purpose; it involves discussion, cooperation and sharing in the carrying out of
fieldwork tasks, and the collection of and collation of data

Negotiation: where students are given the opportunity to contribute to the design,
development and decision making of aspects of the fieldwork activity of which they are part

Risk assessment: there are three stages of risk assessment in fieldwork: (1) the identification
of potential risks or hazards travelling to and from the site(s); (2) the site's potential natural
and human hazards; and (3) potential risks and dangers arising from the activities undertaken.
at the site. For each potential risk or hazard, an assessment must be made of the level of risk,
what needs to be done to mitigate the risk and potential responses to the risk.

Weblinks

Trip intentions

This site provides information about sending your trip intentions to nominated contact people.

PART 1

INTRODUCING GEOGRAPHY

What is geography?

Alaric Maude

Learning objectives

By the end of this chapter, you should be able to:

- understand the five aims of the Australian Curriculum: Geography
- identify the concept or concepts informing a unit and its content descriptions in the Australian Curriculum: Geography
- understand that geography's concepts are complex ideas that have to be unpacked into smaller elements that students can understand and apply
- understand that the concepts are ways of thinking that make the geography curriculum 'geographical', and lead to higher order and powerful thinking

Introduction

The chapter explores the nature of geography as a school subject. It reviews the five aims of the Australian Curriculum: Geography, as these provide a guide for teachers in thinking about their objectives in teaching. It then discusses geography's ways of thinking. These are based on a set of concepts that underpin the curriculum and make it distinctively geographical through the ways in which they view the world, the issues they identify as significant, and the questions, methods of analysis, explanations and criteria for evaluation they generate. These concepts are place, space, environment, interconnection, scale, change and sustainability, and they are unpacked and explained in this chapter.

What is geography?

Some writers define geography by its subject matter. Matthews and Herbert (2008, p. 14), for example, describe it as 'the study of the surface of the Earth. It involves the phenomena and processes of the Earth's natural and human environments and landscapes at local to global scales'. This definition fits the origin of the word 'geography', which comes from the ancient Greek words *geo* and *graphein*, and means 'earth writing'. However, many readers are likely to interpret this to mean that the subject only studies features on the surface of the Earth, such as rivers, landforms, settlements and land use, and the processes that form them, and not intangible things such as economies, communities, migration and opinions. Furthermore, it doesn't distinguish geography from biology, which studies living things on the surface of the Earth, or geology, which studies the physical structure and substance of the Earth, including its surface.

Another approach is to describe geography thematically, as the study of spatial patterns or human–environment relationships. Both themes identify important aspects of the subject, but they are limited in that they each focus on only one of them, either its spatial or human–environment perspective.

If geography cannot readily be defined by its subject matter or its themes, an alternative is to describe it as a way of thinking. Viewing geography in this way was central to the British Geographical Association's influential 2009 manifesto for school geography, *A Different View*. The manifesto argued that 'One way of understanding geography is as a *language* that provides a way of thinking about the world: looking at it, investigating it, perhaps even understanding it in new ways' (Geographical Association 2009, p. 10). Using the analogy of a language, the manifesto identified what it called the *grammar* of geography, which is its concepts and frameworks. This grammar enables students to construct meanings from the subject's large *vocabulary* of factual information, to apply their knowledge to new settings, and to 'think geographically' about the world.

If you are a history teacher assigned to teach geography, you will understand the significance of concepts. One aim of the Australian Curriculum: History is to ensure that students develop an 'understanding and use of historical concepts such as evidence, continuity and change, cause and effect, significance, perspectives, empathy and contestability' (ACARA 2020a). These are ways of interrogating and interpreting the past, and they describe an historical way of thinking that can be applied to a very wide range of subject matter. Similarly, if you are a science teacher you will be familiar with the six key ideas in the science curriculum: pattern, order and organisation; form and function; stability and change; scale and measurement; matter and energy; and systems. These are big ideas that repeat through the science curriculum, and help students to organise their factual knowledge.

The aims of the Australian Curriculum: Geography

While the Australian Curriculum: Geography does not define the subject, it does have a clear statement of the aims of the curriculum. These are to ensure that students develop:

- a sense of wonder, curiosity and respect about places, people, cultures and environments throughout the world

- a deep geographical knowledge of their own locality, Australia, the Asia region and the world
- the ability to think geographically using geographical concepts
- the capacity to be competent, critical and creative users of geographical inquiry methods and skills
- as informed, responsible and active citizens who can contribute to the development of an environmentally and economically sustainable, and socially just world (ACARA 2020b).

The first aim was insisted on by the teachers in the group that drafted the curriculum because it expressed three ideas that they felt were an important part of a geographical education: wonder, curiosity and respect. Wonder is an emotional response to the extraordinary variety, complexity and beauty of the world's places, peoples, cultures and environments. Curiosity is an intellectual response to this wonder, and it makes students want to find out why these places, peoples, cultures and environments are so varied. Respect follows from the knowledge gained through this curiosity, which gives students a growing awareness of the value of these places, peoples, cultures and environments, and why they should be looked after and sustained.

The second aim is to study the world: the places, countries and regions that make up the world, and their peoples, cultures and environments. To do this, the secondary school curriculum does not proceed place by place or country by country, but rather by themes and issues that can be applied to many places. However, each unit also contains a country or world region case study, such as West Asia and North Africa in Year 7, Indonesia in Year 8, a North-East Asian country in Year 9 and India in Year 10; these should be seen as important opportunities to expand the world knowledge of students. All units also provide opportunities to study the local area and other parts of Australia.

The third aim of the curriculum is to develop 'the ability to think geographically, using geographical concepts' (ACARA 2020b), and these are discussed in the present chapter.

The fourth aim is to teach students a range of skills, from observation in the field to statistical and graphical analysis, mapping and spatial analysis, the interpretation of satellite images and the use of geographic information systems. These are explained in Chapters 3–6. These skills are integrated into the process of geographical inquiry, which teaches students how to ask appropriate questions, and then answer them by collecting and analysing information and ideas, drawing and testing conclusions, and communicating results. The inquiry process is explained in Chapter 7.

Students who have developed wonder, curiosity and respect (the first aim), and acquired deep geographical knowledge (the second aim), and geographical thinking and inquiry skills (the third and fourth aims) should have the attitudes, knowledge and skills to be informed, responsible and active citizens, the final aim of the curriculum.

Pause and think

In your teaching of geography, how would you try to achieve the first aim of the Australian Curriculum: Geography?

Geography's concepts

Geography's concepts range from the descriptive, such as city and evapotranspiration, to the more abstract, such as migration and vegetation, to the very abstract, such as space and place. In the Australian Curriculum: Geography, seven of these very abstract concepts have been selected as integral to the development of geographical understanding. These are place, space, environment, interconnection, scale, change and sustainability. The first four have the following characteristics:

- They identify the distinctive ways in which geographers think and describe themes that continually recur in geographical research, such as the interrelationships between people and their biophysical environment (which combines the concepts of environment and interconnection), and the spatial changes that accompany economic development (which is informed by the concept of space).

- 'They are each at the top of a hierarchy of concepts of increasing complexity and abstractness. They synthesise and incorporate simpler and less abstract concepts, and cannot be subsumed by an even bigger and more abstract one' (Maude 2020, p. 234). Such concepts are sometimes termed meta-concepts because they are concepts about concepts, and have to be disaggregated or unpacked if students are to understand what each one means and how to use it. An example is the concept of space, which encompasses subsidiary concepts such as location, distance, spatial distribution and spatial organisation.

- They can 'be applied to a great variety of topics, and across different fields of the subject' (Maude 2020, p. 234).

- They have a number of functions, such as identifying topics worth studying and questions to ask, 'organising information, suggesting methods of analysis, forming generalisations and identifying possible explanations' (Maude 2020, p. 234).

These concepts give the subject coherence, linking the different topics studied through shared concepts and the ways of thinking produced by them. Although they are also employed in 'other disciplines, such as ecology, archaeology, economics and sociology, in none of these are they as central to thinking and practice as in geography' (National Committee for Geographical Sciences 2018, p. 1). This is why it is important for teachers to understand them, and to show students how they are being used in the content they are studying.

The other three concepts – scale, change and sustainability – are equally important but have more limited functions. Scale and change are largely analytical concepts, and are mainly used in geography to explain observations by analysing them at different scales, or over time to see how they have developed, as will be described later in this chapter. Sustainability, on the other hand, is largely an evaluative concept because it is used to assess whether environmental functions, or the economy and population of a place, are being maintained into the future.

The concept of place

The concept of place is about understanding how places are defined and conceptualised, how their characteristics can be explained and the influence they have on our lives and on the outcomes of processes. Places are parts of the Earth's surface that have been identified by people and have meaning for them. They can range in size from a room to a suburb, town or city, region, nation or even the whole planet because these are all areas that have been identified and named by people. However, they may be identified differently by different people. People may also perceive and experience the same places quite differently because age, gender, sexuality, ethnicity or physical disability may make some avoid or feel excluded from places they see as unsafe or hostile.

Places can be described by their characteristics, and these are listed in the Glossary to the Australian Curriculum as including:

> people, climate, production, landforms, built elements of the environment, soils, vegetation, communities, water resources, cultures, mineral resources and landscape. Some characteristics are tangible, for example, rivers and buildings. Others are intangible, for example, scenic quality and socio-economic status. (ACARA 2020c)

All these characteristics constitute the context in which we live, work and play, and in which things happen, and the distinctive characteristics of each place have an influence on our lives and what happens to us.

Explaining why places are like they are

Some of the characteristics of a place are natural, such as landforms, soils and native vegetation. Even these may have been modified by human actions, and many character-istics are described as managed, such as parks, street trees, gardens, cropland and planted forests. Others are constructed, such as buildings and roads. The point here is that places are created by people, and can be changed by people, and this may give students the idea of becoming involved in shaping their own place. This might be through an examination of the liveability of their place, which is the focus of a unit in Year 7, or through an investigation of a proposed new development. However, they should understand that changes promoted by one group may be opposed by others, as is often the case with urban redevelopment projects. They should also understand the constraints on citizen action because of the power of governments, businesses and economic interests, and forces beyond their control. Place-making and place-changing are political processes.

What a place is like is influenced partly by factors internal to it, such as its environment, culture and history. Other influences are external, resulting from its economic, demographic, cultural and political interconnections with other places. Changes in external markets and government policies, or in migration flows, or in cultural influences from outside all produce changes in what a place is like. In explaining the characteristics of a place, we must therefore look both within the place and at its connections with the world. Figure 1.1 shows a place well known to many Australians. What is this place like, and why does it have these characteristics? What are the external influences on this place?

Figure 1.1 Gold Coast, Queensland

How do places influence our lives?

Places can influence people's lives in several ways, all of which can be found somewhere in the Years 7–10 geography curriculum:

- Places provide the services and facilities that support our lives, such as shops, schools, health services, recreational facilities and entertainment venues. Because most people do not travel far to access services and facilities that are required daily or weekly, these need to be provided within or near the places in which people live, but some places are better provided with services and facilities than others, for reasons that students could investigate.

- The places in which people grow up and live may have an influence on their educational attainment and employment opportunities. For example, the subjects available in secondary schools, and the further study and careers to which they lead, vary from place to place, with schools in some places providing only a limited selection.

- People's feelings of attachment to a place or places may contribute to their identity and sense of belonging, and can be important for their wellbeing. For young people in particular, identity is often connected to place.

- Many Aboriginal or Torres Strait Islander people have a very strong attachment to their Country (Aboriginal) or Place (Torres Strait Islands), which is a particular area to which they belong and 'where the spiritual essence of their ancestors remains in the landscape, the sky and the waters' (Queensland Curriculum & Assessment Authority 2020). For Aboriginal or Torres Strait Islander people, this Country or Place has very deep personal meaning based on multiple linkages. These include a strong awareness of the people who lived there before them and from whom they are descended; close relationships with the people living there now; deep knowledge of the environment; and spiritual beliefs about the plants, animals and features of the place and the Dreaming Stories associated with them. People also have obligations as custodians to care for and protect their Country or Place, and the physical and mental wellbeing of Aboriginal and Torres Strait Islander people is often linked to the wellbeing of their Country or Place.

Pause and think

How have the place or places in which you grew up influenced your life?

The effects of place on geographical phenomena and processes

Places are where different environmental and human processes come together and interact, and because each place is unique in its characteristics, the local outcomes of this interaction will differ between places. This is viewing place as an explanatory factor. For example, the globalisation of the economy has not made world cities such as Sydney and Los Angeles the same because of their unique histories, cultures, systems of government and populations. Similarly, no two rivers are the same. Place helps to explain the wonderful variety of the world.

Short-answer questions

1. What is the Aboriginal or Torres Strait Islander name for the Country or Place in which you live or in which your school is located?

2. How would you describe the characteristics of your place? What is different about it compared with other places you know?

3. Are there places that your students try to avoid? If so, why do they try to avoid them?

CONNECTION

Place in the Australian Curriculum: Geography

Place is the organising concept in the units on Place and Liveability, and Sustainable Places.

Key points

- Places are fundamental to human existence because we are always in a place and are consequently always influenced by a place.

- As a concept, place means being aware of the influence of places on people's lives and wellbeing, and on the outcomes of environmental and social processes.

The concept of space

Space can be a complex concept, and academic geographers have developed different ways of conceptualising it. Absolute space is the material space of the surface of the Earth. This space extends in all directions and has no boundaries and no identity, but when parts are named and given meanings, they become places. Positions in this space can be determined geometrically by map coordinates, or latitude and longitude. This is the space of topographical maps, and of maps of territorial units such as property boundaries, administrative areas and planning zones. It is the concept of space most commonly used in physical geography.

Relative space is based not on absolute location and distance, but on the time and cost of moving between places or communicating with people in other places. This depends on the infrastructure built to link places, and will change as infrastructure changes. Because this infrastructure is mainly built by the private and public sectors to increase economic activity by reducing costs and expanding access to markets and resources, it is argued that space is socially produced.

Perceived space, on the other hand, is based on our individual knowledge and experience of places, and how we perceive their location and distance relative to us. This is quite subjective, and will vary from person to person. Perceived space can be explored in schools by asking students to draw a mental map of their city or region. Absolute space is fixed, but relative space and perceived space are constantly changing, and are different for different people and organisations.

Locational concepts

Location is a fundamental element of space, and is described by a number of concepts. *Absolute location* is the unique location of a place as described by latitude and longitude and as shown on topographic maps, while relative location is location in relation to other things, such as the direction and distance of Toowoomba from Brisbane. *Relative location* is much more significant than absolute location, particularly for human phenomena. For example, 'isolated locations distant from major centres are likely to provide fewer opportunities for both businesses (unless they are tied to the location of resources) and individuals than locations in or close to large cities' (Maude 2018, p. 183). Distance is also important in our daily lives because it constrains what we are able to do. For example, we are likely to visit close places more frequently than distant places, simply because of the time and cost involved in travelling longer distances.

The effects of location and distance 'depend on the infrastructure and technology that link places, and the way these are managed by businesses and governments. Improvements in transport and communication systems have greatly reduced the time and cost taken to transport people, goods and information between places, in a process called time-space convergence' (ACARA 2012, p. 5). However, these improvements have been greater in some places than in others. For example, many regions in lower-income countries have seen little benefit from improved communications. Similarly, low-income people are less able to use this infrastructure than higher-income people, and are consequently spatially more constrained.

The following are some other useful locational concepts:

- *Accessibility* is the ease with which people can travel to where employment, shopping, recreation or services such as health are located, and organisations can access the suppliers, services and information they need.

- *Centrality* is the extent to which a location is in the centre of the location of population, customers, businesses and employment. For example, the central business districts (CBDs) of Australian cities have high centrality because of their accessibility from the whole urban area. As a consequence, both land value and building density are very high (see Figure 1.2).

Figure 1.2 The dense development of the central business district of Sydney, a place with a very high degree of centrality. Has the COVID-19 pandemic changed its importance?

- *Proximity* is about closeness to things that are important to a business, organisation or individual people. For some types of business, especially financial and corporate legal firms, face-to-face access to both customers and specialist services is essential, and they tend to cluster together in the centre of major cities.

- *Remoteness* is about places that are relatively far from major population and economic centres. They are likely to have poor access to a range of public and private services.

Spatial distributions

Geographers frequently visualise data spatially (Chapters 3 and 6), and school geography often requires the study of the spatial distributions of a wide range of phenomena. Spatial distributions can also be used to identify possible causal relationships through the method of map comparison. For example, the relationship between vegetation and climate on a global scale can be tested by comparing maps of climatic types and major biomes. A further use is to identify spatial changes over time, such as the spread of a disease, the incidence of drought or the growth of a city, a method that combines the concepts of space and change.

Spatial distributions have patterns or regularities, rather than being random, and these patterns can be analysed for ideas on the causes of the phenomenon being mapped.

To do this, students may need help to identify the patterns. An example is shown in Figure 1.3, which is a map of the distribution of rainfall in Australia. The most obvious feature of the distribution is that rainfall decreases with distance from the coast. However, rainfall extends further inland in the north than in the south, and in the east compared with the west, and higher rainfall areas are a narrow band along the east coast of the continent and the west coast of Tasmania, and a broader band across the north. Each of these elements of the pattern identifies one of the determinants of rainfall in Australia: distance from the coast and sources of atmospheric moisture, elevation (the influence of the Great Dividing Range along the east coast), the tropical monsoon in the north, the subtropical high pressure belt in the centre of the continent and the mid-latitude frontal systems that cross the south from west to east and bring rain in winter.

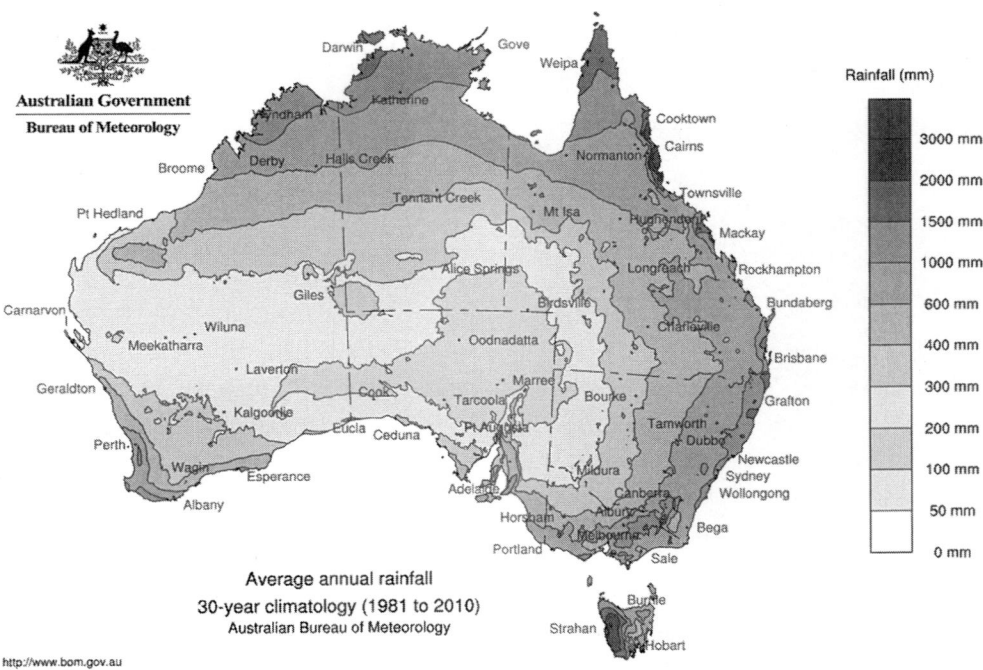

Figure 1.3 Average annual rainfall, Australia

Source: Bureau of Meteorology (2020).

Spatial distributions are not just patterns on a map but have significant environmental, economic, social and political consequences. For example, Australia has the highest percentage of its population living in cities of more than one million people than almost any other country, and this spatial concentration has been a recognised feature for well over a century. Some of the consequences are traffic congestion, suburban sprawl, air and water pollution, water supply problems and the dominance of Australian politics by the interests of people in the capital cities. It is therefore not enough to ask students to explain a spatial pattern; they should also think about the effects of that pattern. In explaining a pattern, they should also be made aware of the relationships between spatial distributions and society. For example, the spatial pattern of advantaged and disadvantaged social groups found in most cities is both a result of the social inequalities in that society and a contributor to them through the effects of place on people described earlier. Space and society are interrelated, and spatial patterns both result from and reproduce the social and economic structures of societies.

The organisation of space

Societies divide up space for a variety of purposes, such as creating boundaries for states and local government areas, defining the catchment areas of schools, and creating areal divisions in organisations such as Rotary. These are all examples of the organisation of space.

CONNECTION

Space in the Australian Curriculum: Geography

Space is the organising concept in the units on Changing Nations and Geographies of Human Wellbeing.

Key point

- The concept of space teaches us to think spatially by recognising the influence of location and distance on people and places; using spatial distributions, patterns and trends to identify causes and consequences; recognising the interrelationships between space and society; and understanding how people perceive, organise, manage and use space.

The concept of environment

The term 'environment' can have a range of meanings in geography, but in the Australian Curriculum: Geography it refers to the physical and biological environment – the living and non-living elements of the Earth's surface and atmosphere, including those created by people such as croplands, planted forests, buildings and roads. In secondary school, the more precise term 'biophysical environment' is sometimes used. Geographers view this environment as the home of humankind, so the geographical concept of environment is about the significance of the environment to people and other living things, how to understand how it works, and how to conceptualise and understand the interrelationships between people and environment. These are all ways of thinking about the environment.

The significance of the environment for people

The significance of the biophysical environment for people can be described by the four environmental functions that support human life and wellbeing. These are listed in the Glossary to the Australian Curriculum: Geography, and are shown in Figure 1.4.

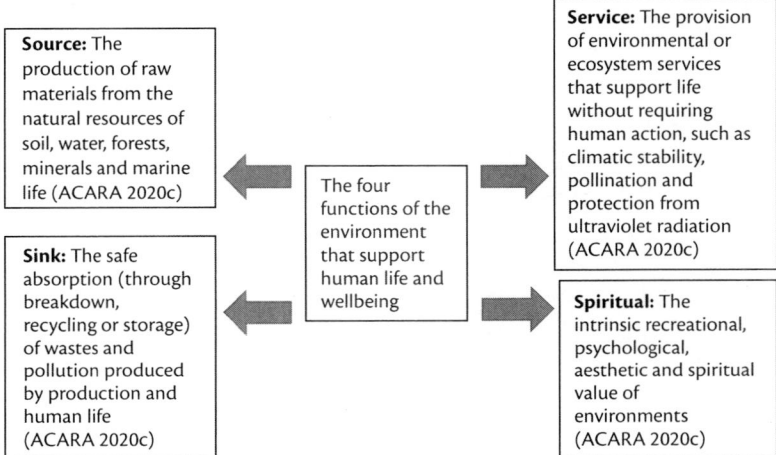

Source: The production of raw materials from the natural resources of soil, water, forests, minerals and marine life (ACARA 2020c)

Sink: The safe absorption (through breakdown, recycling or storage) of wastes and pollution produced by production and human life (ACARA 2020c)

The four functions of the environment that support human life and wellbeing

Service: The provision of environmental or ecosystem services that support life without requiring human action, such as climatic stability, pollination and protection from ultraviolet radiation (ACARA 2020c)

Spiritual: The intrinsic recreational, psychological, aesthetic and spiritual value of environments (ACARA 2020c)

Figure 1.4 The four functions of the environment for people

These functions range from the 'practical (such as the provision of food and water) to the emotional (such as inspiring landscapes)' (Maude 2018, p. 182). Note that the term 'spiritual' is meant not just in a religious sense, but to include everything from people's emotional feelings about particular environments to the deeply spiritual relationships of Aboriginal and Torres Strait Islander peoples with their Country or Place. For many people, the environment is far more than a resource to be used, and is also a source of enjoyment and inspiration, while for some it is an integral part of their spiritual life, wellbeing and identity (Figure 1.5).

Figure 1.5 Two iconic Australian landscapes, one inland (left, Tjoritja/West MacDonnell National Park, Northern Territory) and one coastal (right, Byron Bay, New South Wales). Do these represent many people's sense of their Australian identity?

Understanding the environment

The biophysical environment is the product of the interactions between the atmosphere, oceans, landforms, geology, vegetation, animals and people. Geographers are particularly interested in the ways these elements vary spatially across the Earth's surface, and are integrated in places to create distinctive environments. This is the large and important field of physical geography.

Understanding the environment, and particularly the ways in which it is changing, also requires an understanding of the effects of people on that environment because humans have altered environments over thousands of years, and are now doing so at an accelerating rate. This has had both positive and negative outcomes. The positive outcomes include increased food, fibre and wood production, income and materials from mining, the ability to store water and the prevention of flooding. On the other hand, human mismanagement of the environment can have negative effects if processes such as soil erosion, air and water pollution, vegetation removal and climate change reduce the capacity of the environment to support human life through the four functions. Degradation of the environment also reduces the quality of human life and can threaten human health, while vegetation clearance reduces the habitat available for animal life and contributes to climate change.

The influence of the environment on people

The biophysical environment presents both opportunities for, and constraints on, human settlement and economic activity. Opportunities include agricultural, mineral and marine resources, topography that facilitates transportation, and environments that attract tourists. Constraints include limited rainfall, rugged topography, poor soils, lack of mineral resources and extremes of heat or cold. These can be reduced but not eliminated by technology and human organisation, partly because technological solutions often produce new problems.

Geography has a long history of debate over the extent of environmental influence on human economies and societies. While some past ideas have been discredited (such as environmental determinism, which contended that the economies and even cultures of societies were determined by their biophysical environment), contemporary examples of environmental influences include that of climate on population distribution in Australia, rainfall on agriculture, minerals or water resources on regional development and landscape on Australian identity. However, teachers need to be careful when suggesting that the environment causes something, as the reality is usually more complex.

The influence of the environment on people depends on how they perceive, adapt to and use it. For example, a low-rainfall environment could be used for hunting and gathering, grazing of cattle for sale, growing crops with irrigation water, extracting minerals or natural gas, for biodiversity conservation and nature tourism, or the location of a city such as Las Vegas. The use will depend on culture, population density, type of economy, human organisation, level of technology and other factors.

The environment as hazard

Different environments have different natural hazards, but as their impact on people is determined by human as well as environmental factors, they are not purely 'natural'. For example, the drainage of wetlands removes areas that once absorbed floodwaters, and consequently may increase damage from flooding. Humans may also be having an influence on natural hazards if global warming is increasing the frequency and/or severity of hazard events.

Pause and think

Have your views of our ability to manage the environment changed as a result of the 2020 coronavirus pandemic?

CONNECTION

Environment in the Australian Curriculum: Geography

Environment is the key organising concept in the units on Water in the World; Landforms and Landscapes; Biomes and Food Security; Environmental Change and Management; Natural and Ecological Hazards; and Land Cover Transformation.

Key points

- The concept of environment draws attention to the influences of the environment on people, and of people on the environment.

- Humans are completely 'dependent on the environment for their survival and wellbeing' (Maude 2020, p. 236).

- It is consequently vital to understand how the environment works, and how and why it is changing.

The concept of interconnection

The concept of interconnection is about 'recognising that nothing studied in geography exists in isolation because everything is influenced by its relationships with other phenomena, both within and between places' (Maude 2020, p. 236). It is an important part of geographical thinking, and is the opposite of the reductionism common in the natural sciences and in some of the social sciences, in which the focus is often on a very narrow set of variables. In geography, the idea of interconnection is expressed through a number of subsidiary concepts.

Interdependence

Interdependence is a form of interconnection. It is about the mutual dependence of phenomena on each other, as in biological systems where organisms are dependent on each other for their survival. In human geography, it is usually about how people in places that are connected by trade are interdependent on each other, and it is frequently linked with aspirations such as fair trade and global citizenship.

Causal relationships

Causal relationships are the result of interconnections between two or more phenomena. From a geographical viewpoint, these could be thought of as two types:

1. Causal relationships *within* the one place. These are the interconnections between:

    – the environmental characteristics of a place, such as the effects of climate on vegetation and vegetation on climate; these relationships are fundamental to much of physical geography

    – the human characteristics of a place, such as the influence of the economic structure of a town on the characteristics of its population, and of the characteristics of its population (such as skills) on its ability to develop new industries

    – the environmental and human characteristics of a place, such as the influence of the environment on agriculture or of people's actions on the environment. In descriptions of geography's concepts this is often the only type of interconnection discussed.

2. Causal relationships *between* places, such as the effects of external demand on the economy of a place, or of upstream irrigation on downstream water quantity and quality.

Processes

Here, a process means a sequence of connected events. An example is erosion and sedimentation, which combine to form a process in which weathered material is transported by water or wind and deposited in a new place through a sequence of connected events. Chain migration is another example, in which the initial movement of individuals to a new place may, through the flow of information back to the place of origin, encourage further migration. In this example, the places are interconnected through flows of people, information and probably remittances of money.

Flows

Flows are things such as the movement of water in a river or migration between cities. They have an effect on the places connected by these flows, so to understand places and why they are changing we have to understand the flows between them.

Systems

Systems are sets of interconnected objects linked through flows of energy and matter (Maude 2020, p. 236) and, where humans are involved, people and information. They are a useful way to understand change because change in one element of the system will produce change in other elements due to the interconnections between them. An example is dryland salinisation.

Holistic thinking

An awareness of interconnections encourages holistic thinking. This is the ability to look for the relationships between the phenomena studied and other phenomena, to be aware of multiple possible explanations for our findings and to draw ideas from both the natural and social worlds.

Holistic thinking is important when trying to understand environmental change. The Connection box on dryland salinisation identified the environmental changes that resulted from clearing vegetation. But why was the vegetation cleared? Were farmers unaware of what would happen because it was not a problem they had encountered before? The history of salinity problems in the Wheatbelt of Western Australia shows that this is not an adequate explanation because the cause of salinity was identified in the 1930s, and clearing continued for decades. Was clearing encouraged by governments fixed on 'development'? Was it demanded by banks as a condition for a loan? Was it a response to declining prices for farm output, combined with increased costs of production, putting pressure on farmers to increase output? To understand why environmental degradation occurs, and therefore how to prevent it, we must understand the environmental processes producing the degradation (as shown in Figure 1.6); the human actions that have initiated these processes (vegetation clearance); and the attitudinal, demographic, social, economic and political causes of these human actions. All these questions can be analysed through a human–environment system framework that integrates the specific environmental, social, economic and political elements that are interconnected in the problem being analysed.

Pause and think

Identify an environmental problem and use holistic thinking to work out what might be causing the problem.

CONNECTION

Dryland salinisation

An example of the application of systems thinking that is very relevant to the Australian Curriculum: Geography is dryland salinisation. Figure 1.6 shows the changes that can occur when native vegetation is cleared in places where the groundwater is naturally saline. The situation can be described as a system because of the interconnections between vegetation, evapotranspiration, the volume of water penetrating into the soil and the level of the water table. When the native vegetation, with its deep roots, is removed, all the other variables in the system change.

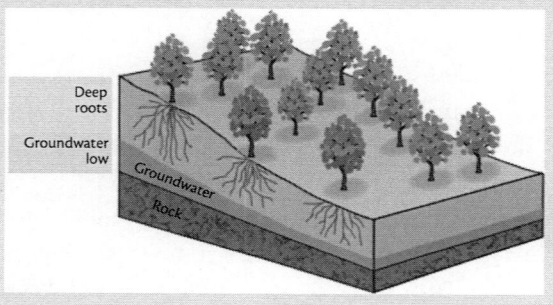

Trees, deep-rooted perennials and native vegetation use most of the water that enters the soil, and through evapotranspiration return much of it back to the atmosphere. The result is limited water flow below the plant root zone and a relatively deep and limited layer of groundwater.

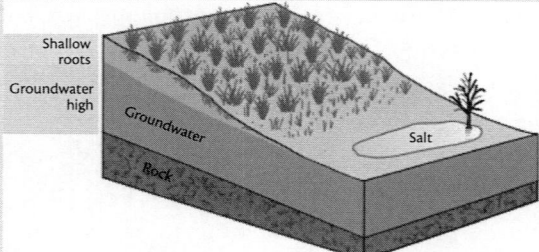

Removing the native vegetation and replacing it with shallow-rooted annuals reduces evapotranspiration into the atmosphere, and increases the flow of water into the groundwater. The result is that the water table rises, bringing salt to the root zone and the soil surface if there is salt in the environment.

Figure 1.6 The effects of vegetation clearance on the water table

Source: Adapted from Walker, Gilfedder & Williams (1999).

Question

Why would farmers clear vegetation if this is going to be the result?

CONNECTION

Interconnection in the Australian Curriculum: Geography

Interconnection is the key organising concept in the Geographies of Interconnections and Global Transformations units, and is also important in all other units.

Key point

- The concept of interconnection prompts us to look outside whatever we are investigating and at its linkages with other phenomena, and to think holistically about causes and effects.

The concept of scale

Scale here refers to the size of the spatial units being analysed, and is different to scale on a map, which is the relationship between distance on the map to distance on the ground. The first can be called observational scale, and the second can be termed map or cartographic scale. Observational scale is an analytical concept because it is about looking for and testing relationships at different scales. The scale used in analysis could be areas as small as 1 square metre, which might be appropriate when studying the distribution of plant species, or as large as the whole world, appropriate for studying the distribution of biomes.

Analytically, scale can be used in several ways. One is to identify whether measures at a national scale conceal significant variations at regional or local scales. Geographers frequently describe scale as being like using a zoom lens, in which whatever is being studied is viewed at different scales. For example, if life expectancy is examined at a global scale, Australia ranks in the top 10 countries in the world, with a life expectancy for men in 2017–19 of 80.9 years (ABS 2020). At the scale of the states and territories the lowest life expectancy is 75.5 years in the Northern Territory (ABS 2020), which is similar to countries such as Morocco. At the scale of ABS statistical areas, the lowest figure is 72.2 years in the Northern Territory Outback (ABS 2020), which is similar to countries such as Thailand and Sri Lanka. Inequalities in health in Australia become more and more apparent as we zoom into smaller and smaller areas.

Another analytical use of scale is to test whether relationships found at one scale may be different at other scales. For example, in studies of vegetation, climate is the main influence at the global scale but soil and drainage may be dominant at the local scale. A third application of scale is to look for causal relationships that cross scales. For example, local events can have global outcomes, such as the effects of local vegetation removal on global climate, or the collapse of the New York Stock Exchange in 2007–08 on the global economy. Similarly, global events, such as climate change, have local impacts that differ from region to region.

CONNECTION

Scale in the Australian Curriculum: Geography

Scale is a concept that can be applied in all units. It is a major organising concept in Geographies of Human Wellbeing, in which wellbeing is examined at global, national and local scales.

Key point

- Scale is an analytical concept used to choose the most appropriate spatial units for analysis, and to test causal relationships.

The concept of change

The concept of change is about explaining geographical phenomena by finding out how they have developed over time. The evolution of landforms can help to explain their characteristics, while the historical development of cities and towns helps to explain the factors that have caused their growth or decline. In both cases, what the landform or urban area is like now is influenced by what it was like in the past. Understanding how things have evolved over time is especially valuable if it can be used to project change into the future. For a town, this could mean determining how it has changed over several decades, identifying the important trends in business and population, and projecting them forward. Students can then evaluate the result and, if they don't like it, think about what would change the trends that were responsible for the projected outcome.

Changes in spatial distributions can be used as geographical indicators of environmental, economic, demographic, political and other changes. The Year 8 Changing Nations unit, for example, examines development in low- and middle-income countries through the perspective of urbanisation, which is a profound change in the spatial distribution of population, whereas an economist might study development through changes in the structure of the economy. Another example might be to use changes in the location and extent of areas burned by bushfires as a possible indicator of climate change.

CONNECTION

Change in the Australian Curriculum: Geography

Change is a concept in all the units.

Key point

- The concept of change is about gaining an understanding of things by finding out how they have changed over time and how they might change in the future.

The concept of sustainability

What is sustainability?

Sustainability can be tricky concept, as different writers have different definitions, and in Australian school textbooks it is sometimes equated with sustainable development. The word 'sustainability' is

> a noun formed from the adjective sustainable, which means being able to be maintained or kept going, so something is sustainable if it can be kept going into the future ... Sustainability, then, is the state or condition of being sustainable, whereas sustainable development is a process of economic and social change designed to produce an environmentally sustainable economy and a just society. (Maude 2014, p. 19)

Here we are interested in sustainability as a condition because the concept can then be used to evaluate whether or not something is sustainable. A suggested definition for environmental sustainability is:

> Environmental sustainability is the maintenance of the capacity of the biophysical environment to support human life and wellbeing into the future.

Pause and think

This definition of sustainability does not mention non-human life, such as animals. Should it? If so, why and how?

This general statement can be made more specific if 'the capacity of the biophysical environment' is disaggregated into the source, sink, service and spiritual functions of the environment described earlier in the chapter. For each of these, we can 'describe what sustainability means for that function' and so provide students with 'criteria for assessing sustainability in different situations' (Maude 2014, p. 20).

Environmental resources (the source function) are either renewable (e.g. water) or non-renewable (e.g. minerals), and there is a separate sustainability principle for each. Renewable resources are sustainable if they are being extracted at or below their rates of renewal, and in ways that do not reduce the productive capacity of the environment. For example, a water resource, such as a river or groundwater, can be used sustainably (i.e. into the future), provided that water is not extracted faster than it is being replenished by rainfall or groundwater inflow. Otherwise the river dries up or the water table drops. However, the sustainability of a renewable resource is also compromised if the methods of extracting the resource reduce the productive capacity of the environment, which is the second part of the principle. This can occur if water resources become polluted, or agricultural land is degraded through soil erosion, compaction, salinisation and acidification.

Non-renewable resources are sustainable if they are extracted no faster than the rate at which they can be maintained through the discovery of new reserves, recycling or substitution, and in ways that do not damage other environmental functions. The first part of this principle is often ensured by market forces because when a non-renewable resource becomes scarce its price rises, and more effort is made to locate new reserves, recycle the resource, and/or develop substitutes. However, sustainability also depends on the extraction of resources not damaging other environmental functions, which can occur if, for example, mining damages the quality of underground water for agriculture, or the extraction and burning of fossil fuels produce climate change.

Pause and think

Can a non-renewable resource be converted into a renewable resource?

This may seem a logical impossibility, but if some of the income from exploiting the resource is converted into financial capital and invested, it becomes a renewable resource. Countries such as Norway have done this with their oil revenue, and built up large financial reserves. Australia has not.

Wastes (the sink function of the environment) are either biodegradable or non-biodegradable. For biodegradable wastes, the sink function of the environment is sustainable as long as they are not added to the environment in ways that prevent them from being broken down and safely recycled or stored, or reduce the productive capacity of the environment, or threaten human health. For non-biodegradable wastes, the sink function of the environment is sustainable as long as these are not added to the environment in quantities that threaten human health or other environmental functions. If this principle is not followed, the source function of the environment could be threatened by algal blooms, or by

> toxic chemicals that make land or water resources unusable, or kill marine and aquatic life. Human health could also be threatened. This principle is the subject of a growing body of environmental regulation which bans or limits the discharge of toxic substances into the environment, with the aim of keeping their concentrations below the levels that affect the health of humans and other living bodies. (Maude 2014, p. 21)

The third function of the environment is the provision of ecosystem services, which are the free environmental services that support life without requiring human action. They are frequently overlooked and taken for granted but are of crucial importance in maintaining human life and human wellbeing, and have considerable economic value. Sustainability requires the protection of the ecosystem services function of the environment.

The fourth function of the environment is its aesthetic, emotional and spiritual value to people. Consequently, sustainability requires the protection of these values. This principle differs from the others in that it cannot be measured objectively because it is about human perceptions, feelings, beliefs, values and worldviews, and these differ from person to person. How to apply the principle is therefore a matter of continuing debate and sometimes conflict between different groups, and in the end is determined politically.

It is important to note that sustainability is not about preventing development and environmental change:

> Humans have a long history of transforming their environments in ways that have enabled the world to support more people at higher levels of wellbeing. If humans had remained hunter-gatherers and not cleared land for farming, the world might support about only 100 million people, and would not have developed urban civilisations. If modern agricultural technologies had not been developed the world might support only about three billion people, instead of the present nearly eight billion. Our way of life has been made possible by changes to the environment. Sustainability is about ensuring that these changes do not threaten important environmental functions, through actions like carefully managing renewable resources, preventing soil degradation, restricting climate change, and preserving genetic resources. (Maude 2014, p. 23)

Disagreements about sustainability

Sustainability can be a sensitive topic to teach because opinions on its significance and how to manage it are often strongly divided. This can be an advantage for teachers because it can stimulate student debate over why people differ and how to judge between opposing opinions.

At least two types of disagreement can be identified. First, for all four functions of the environment there can be disagreement over whether a situation is serious enough to require action. Some people even deny that a problem exists, despite the scientific evidence. This attitude is often based on a dislike of government regulation of business, and a belief that economic and social considerations, like the generation of income and employment, should have precedence over the environment. Others may exaggerate problems, and seize on sustainability issues to argue for greater regulation, or changes in our economy and ways of living, because of a dislike of capitalism and consumerism.

Second, there are different views on how to achieve sustainability – for example, to ensure the sustainability of a renewable resource some prefer methods that reduce the use of the resource, such as restrictions on fishing or more recycling of paper and cardboard to reduce the need to harvest new wood. Others want methods that increase the output of the resource, such as farming fish (aquaculture) and planting trees.

Using the concept

This section has set out a way of defining sustainability that students can use to evaluate different situations, and to turn what is often a very vague idea into something that can be made operational. In teaching, it can also be an integrating idea because understanding the causes of unsustainability requires an exploration of both environmental and human factors, as described earlier in the section on holistic thinking, and to develop acceptable programs to improve sustainability requires a balancing of environmental, economic and social considerations.

Short-answer questions

1. The Glossary to the Australian Curriculum: Geography defines sustainability as an 'ongoing capacity of an environment to maintain all life, whereby the needs of the present are met without compromising the ability of future generations to meet their needs' (ACARA 2020c). In what ways does this definition differ from the one at the beginning of the section? Which do you prefer, and why?

2. Find a local sustainability issue, such as declining fish stocks, air pollution, water shortages or a proposal to clear native vegetation for agriculture. Investigate the issue as a class and debate what should be done.

Place sustainability

The concept of sustainability can also be applied to a place, and this is the focus of the Year 11 unit on Sustainable Places. This unit gives students the opportunity to examine the interconnected economic, demographic, social and environmental challenges of a particular place, with a focus on issues such as 'population growth and decline, employment, economic restructuring, transport infrastructure needs, housing, demands for improved health and education services, and other matters related to liveability' (ACARA 2020d).

CONNECTION

Sustainability in the Australian Curriculum: Geography

Sustainability is an important concept in Biomes and Food Security, Environmental Change and Management, and Sustainable Places.

Key points

- Sustainability is about maintaining the source, sink, service and spiritual functions of the environment that support human life and wellbeing.

- Sustainability can also be applied to places, where it becomes much more than just environmental sustainability.

Why the concepts are important

The seven concepts explained in this chapter have a number of functions in school geography.

They make the subject 'geographical'

The seven concepts, and the more specific ideas into which they can be disaggregated, are what make the geography curriculum distinctively 'geographical' because they underly the questions that geographers ask, and the ways they organise, explain, analyse, generalise and evaluate. As ways of thinking they describe how geographers look at the world, and by being repeated throughout the school curriculum, they give the subject strong coherence.

The concepts also enable students to 'distinguish what is distinctively different about learning geography from other school subject areas' (Brooks 2018, p. 106). Consequently, it is important that students understand the conceptual foundations of the units they are studying, and what makes the way those units are being studied 'geographical'. An Australian Academy of Science report on Australian geography argues that:

> Geography is distinctive in its emphasis on spatial thinking, its interest in knowledge generated from the study of specific places, and its recognition of the fundamental importance of the environment to human welfare. Its vision is both local and global. It is also marked by an awareness of the interconnections between phenomena and processes both within places and across space, and its fields of study span the natural sciences, social sciences and humanities. In a world in which inequalities within and between places can threaten social cohesion, the pressure of human impacts on the environment is a growing concern, places and people are increasingly interconnected globally, and problems require answers that integrate different fields of knowledge, Geography has much to offer. (National Committee for Geographical Sciences, 2018, p. 4)

They link different units conceptually

Making the conceptual foundations of the units apparent will also help students to see geography as a coherent subject, rather than just a succession of unrelated topics, because the same concepts will recur in different units. For example, while the concept of space underlies the unit on Changing Nations, it is also the focus of the unit on Geographies of Human Wellbeing. Similarly, the idea of human impact on the environment, which is part of the overall concept of environment, is an important component of the units on Landforms and Landscapes, and Biomes and Food Security, and a subsidiary concept in Place and Liveability and Geographies of Interconnections.

They can be used to ask geographical questions

Asking good questions is an important way of learning, and the concepts can be used to make these questions geographical ones. An example might be: Have the ways in which people have altered this river system changed the risk of flooding? This question is based on the concepts of environment and change. Another example is: Why is the Australian population more concentrated in cities of greater than one million than in the United States? This question is based on the concepts of space and place (the latter because the question compares two places, Australia and the United States).

They suggest ways of organising data

Data that are recorded and portrayed by place, such as by local government areas, exemplify the concept of place, while data that are mapped exemplify the concept of space. These are two very common ways of organising and portraying data in geography.

They underly geographical methods of analysis

Analysis is the breaking down of a complex problem into its different parts and studying each of these separately. Specifically geographical methods of analysis include comparisons between places (concept of place) and comparisons of spatial distributions (concept of space), and for many students these will represent new ways of thinking. Geospatial technologies allow students to undertake some quite sophisticated spatial analysis, using readily available software.

They are geographical ways of explaining

Explaining is finding the causal relationship(s) that account for why something is the way it is, or why it is changing. This means being able to explain the processes or mechanisms by which a cause produces an effect. A geographical explanation might relate to the distinctive characteristics of a place (concept of place), relative location (concept of space) or the result of change in one element in an interconnected system (concept of interconnection).

They enable students to generalise

A generalisation is a statement of the relationship between two or more concepts derived from a case study. Generalisations are educationally valuable because they help students to summarise a lot of information, and thus increase their ability to retain it. More importantly, they enable them 'to apply what they have learned from the study of one set of relationships to new situations that they have not encountered before' (Maude & Caldis 2019, p. 34). This enables them to 'ask appropriate questions and make sense of contexts beyond their experience' (Maude & Caldis 2019, p. 34), and so create new understandings or solve new problems. An example of a geographical generalisation is this statement:

> Coastal areas are dominated by wave and tidal processes that drive weathering . . . [and] sediment movement, and stopping natural sediment movements in one location on the coast may cause additional erosion and major coastal problems [elsewhere]. (Adapted from Holden 2011, p. 119)

This statement combines the environmental concepts of 'wave processes, tidal processes, weathering, erosion and sediment movement, and links them together through the concept of interconnection. It also involves the concept of place, because it identifies that actions taken in one coastal place may have impacts in another' (Maude 2020, p. 238). The generalisation can be applied to make sense of the 'dynamics of a wide variety of coasts, and also alerts students to the potentially negative effects of structures such as marinas, and the importance of locating them in places where sediment movement is minimal' (Maude 2020, p. 238).

A generalisation is probably the nearest that geographers can get to a principle or law; however, because of the uniqueness of places, geographical generalisations may have many exceptions – particularly when human actions are involved. Such cases can be used to get students thinking about how to explain why the case is different, and whether it negates the generalisation, which will involve them in some serious thinking.

The concept of sustainability can be used to evaluate

The concept of sustainability can be applied to evaluate whether human impacts on the environment are threatening the sustainability of one or more of the four environmental functions that support human life and wellbeing.

Readers will recognise that these examples of ways of using the concepts are also forms of higher order and critical thinking. Higher order thinking occurs 'when students go beyond recalling [factual] knowledge to demonstrate their understanding of it in increasingly complex and creative ways' (Maude & Caldis 2019, p. 30). These include analysing and synthesising information, finding answers to questions, and transferring their understanding to solve problems in contexts that have not been encountered before. Critical thinking overlaps with higher order thinking, and is about evaluating evidence and claims to knowledge, using reasoning to find answers to questions and problems and reflecting on these answers. The term 'critical' does not mean that it is only concerned with criticising other people's arguments, as it should be applied to one's own thinking, and then followed by creative thinking to find better answers. It could be thought of as the application of higher order thinking skills to 'interpret, analyse and evaluate ideas and arguments' (Fisher 2011, p. v).

Recent graduates may also recognise that the examples above are forms of powerful knowledge. This concept was introduced into educational thinking over a decade ago by Michael Young, a British sociologist of education, and has been picked up by geography educators in several countries (Brooks, Butt & Fargher 2017). Young contends that 'the main purpose of schools is to teach knowledge that enables students to understand and think beyond the limits of their own experience, and describes such knowledge as powerful' (Maude & Caldis 2019, p. 32). Geography's concepts can be powerful in this sense if they enable young people to:

- discover new ways of thinking, such as those described in this chapter

- explain and understand the natural and social worlds

- use generalisations to analyse and understand situations they have never experienced themselves

- use the concept of change to forecast futures and what they could do to influence them, and

- use the concept of sustainability to evaluate environmental changes (Maude 2015).

Conclusion

This chapter has described the five aims of the Australian Curriculum: Geography, objectives that teachers should keep in mind in designing and teaching units of study. It also suggested that geography may best be described as a way of thinking about a very wide range of subject matter, based on the concepts of place, space, environment, interconnection, scale, change and sustainability. These concepts were unpacked, explained and illustrated, and should be explicitly incorporated into teaching, learning and assessment. They develop a geographical way of thinking, deepen students' understanding of the subject, give the subject coherence across the different units in the curriculum, and involve students in higher order and powerful thinking.

Bringing it together

1. How might your understanding of the aims of the Australian Curriculum: Geography influence how you teach the subject?

2. What are the functions of the seven geographical concepts in the curriculum?

3. How can generalisations be used?

4. Choose one of the units in the Australian Curriculum: Geography for Years 7–12. How are the concepts illustrated in the Knowledge and Understanding strand? Is there an overall concept that is shaping the unit?

5. How would you teach a unit in a way that requires students to engage in powerful forms of geographical thinking?

References

Australian Bureau of Statistics [ABS] (2020) *Life tables: Key statistics*, Canberra: ABS, retrieved from www.abs.gov.au/statistics/people/population/life-tables/latest-release

Australian Curriculum, Assessment and Reporting Authority [ACARA] (2012) *Revised Draft Foundation to Year 12 Australian Curriculum: Geography*, Canberra: ACARA, retrieved from https://docs.acara.edu.au/resources/Draft_F-12_Australian_Curriculum_-_Geography_Validation_Version_29082012.pdf

——(2020a) *Australian Curriculum: History aims*, Canberra: ACARA, retrieved from www.australiancurriculum.edu.au/f-10-curriculum/humanities-and-social-sciences/history/aims

——(2020b) *Australian Curriculum: Geography aims*, Canberra: ACARA, retrieved from www.australiancurriculum.edu.au/f-10-curriculum/humanities-and-social-sciences/geography/aims

——(2020c) *Australian Curriculum: Geography glossary*, Canberra: ACARA, retrieved from www.australiancurriculum.edu.au/f-10-curriculum/humanities-and-social-sciences/geography/glossary

——(2020d) *Australian Curriculum: Geography Senior Secondary Unit 2 Sustainable Places*, Canberra: ACARA, retrieved from www.australiancurriculum.edu.au/senior-secondary-curriculum/humanities-and-social-sciences/geography

Brooks, C. (2018) Understanding conceptual development in school geography, in M. Jones & D. Lambert, eds, *Debates in geography education*, 2nd ed., Abingdon: Routledge, pp. 103–14.

Brooks, C., Butt, G. & Fargher, M. (eds) (2017) *The power of geographical thinking*, Cham: Springer.

Bureau of Meteorology (2020) *Average annual rainfall – 30-year climatology (1981 to 2010)*, Canberra: Australian Bureau of Meteorology and Commonwealth of Australia, retrieved from www.bom.gov.au/jsp/ncc/climate_averages/rainfall/index.jsp

Fisher, A. (2011) *Critical thinking: An introduction*, 2nd ed., Cambridge: Cambridge University Press.

Geographical Association (2009) *A different view: A manifesto from the Geographical Association*, Sheffield: Geographical Association.

Holden, J. (2011) *Physical geography: The basics*, London: Routledge.

Matthews, J.A. & Herbert, D.T. (2008) *Geography: A very short introduction*, Oxford: Oxford University Press.

Maude, A. (2015) What is powerful knowledge and can it be found in the Australian geography curriculum?, *Geographical Education*, 28, 18–26.

——(2018) Geography and powerful knowledge: A contribution to the debate, *International Research in Geographical and Environmental Education*, 27(2), 179–90.

——(2020) The role of geography's concepts and powerful knowledge in a Future 3 curriculum, *International Research in Geographical and Environmental Education*, 29(3), 232–44.

Maude, A. & Caldis, S. (2019) Teaching higher-order thinking and powerful geographical knowledge through the Stage 5 Biomes and Food Security Unit, *Geographical Education*, 32, 30–9.

National Committee for Geographical Sciences (2018) *Geography: Shaping Australia's future*, Canberra: Australian Academy of Science, retrieved from www.science.org.au/files/userfiles/support/reports-and-plans/2018/geography-decadal-plan.pdf

Queensland Curriculum & Assessment Authority (2020) Relationship to Country/Place, retrieved from www.qcaa.qld.edu.au/about/k-12-policies/aboriginal-torres-strait-islander-perspectives/resources/relationships-place

Walker, G., Gilfedder, M. & Williams, J. (1999) *Effectiveness of current farming systems in the control of dryland salinity*, Canberra: CSIRO.

CHAPTER
2

Core knowledge through case studies

John Butler

Learning objectives

By the end of this chapter, you should be able to:

- identify the curriculum pattern within the Australian Curriculum: Geography
- understand how to interpret the key ideas and core knowledge in each of the units of the Australian Curriculum: Geography
- consider how to use case studies to develop core knowledge and concepts with the Australian Curriculum: Geography

Introduction

Every school subject and every field of knowledge has a similar problem: What is the core knowledge that should be learned? In university courses, each student takes a different pathway through the options that are offered, and in school courses teachers and curriculum planners have to choose which ideas, content and skills should be emphasised. No student or teacher will ever have the time to comprehensively study every area of their major subjects; in fact, the longer they study a subject at university, the more they tend to increase the depth of their knowledge rather than the breadth.

The Australian Curriculum: Geography has been written to include a spread of geographical content, with topics suited to the age levels of students. Teachers are given a large amount of information about each topic to help them plan their teaching. It is the task of the teacher to understand what the Australian Curriculum says about each of the key elements of curriculum, listed in the sections below, and to decide how to integrate them into the teaching program.

This chapter aims to help teachers unpack the large amount of information in the Australian Curriculum: Geography, emphasising the core knowledge and understandings. For every unit of the curriculum from Year 7 to Year 10, as well as the senior years, an interpretation of the core knowledge is presented, together with the key inquiry and skills developed within it. Following this are commentaries on examples of appropriate case studies that are suitable to achieve the aims of the unit.

The Australian Curriculum: Geography

Pause and think

What areas of geography have you personally studied in your university courses? Which areas of geography have you never studied at tertiary level? What influence will these personal characteristics have on your teaching and interpretation of the curriculum?

The key elements of the Australian Curriculum: Geography are:

- inquiry questions
- geographical knowledge – content descriptors
- geographical inquiry and skills – content descriptors
- achievement standards (for assessment)
- methods and techniques of teaching.

The first four of these are explicitly stated in the Australian Curriculum: Geography. The fifth element (methods and techniques of teaching) is decided by each individual teacher or school faculty. The *inquiry questions* give a teacher an overall direction to aim for. The *achievement standards* are a summary of the learning that should be achieved within the topic. The *content descriptors* give much more detail about the learning that is intended. For each content descriptor, there are a number of elaborations.

Teachers should be very clear about the difference between the content descriptors and the elaborations. Elaborations are not compulsory – they are suggestions of case studies, examples, techniques and comparisons. These are all helpful for a teacher planning their program. The commentaries on examples of case studies should be treated as examples only: they are suggestions of ways of combining content, skills and concepts within case studies.

Pause and think

The Australian Curriculum deliberately does not specify methodology of teaching. However, there are certain indications in content, skills, concepts and priorities that imply or support certain methodologies. What are some of these? What do they suggest to you?

Core knowledge of students at the end of Year 6

Before looking in detail at the core knowledge in the Australian Curriculum: Geography from Years 7 to 10, it is worthwhile looking at an overview of what you can expect a student who has just finished Year 6 to have learned.

CONNECTION

Australian Curriculum: Geography F–6

The inquiry questions for geography from Foundation to Grade 6 present an overview of the learning expected in those year levels. Read through each year level and answer the questions that follow.

Foundation Year: People live in places

Inquiry questions

- What are places like?

- What makes a place special?

- How can we look after the places we live in?

Year 1: Places have distinctive features

Inquiry questions

- What are the different features of places?

- How can we care for places?

- How have the features of places changed?

Year 2: People are connected to many places

Inquiry questions

- What is a place?

- How are people connected to their place and other places?

- What factors affect my connection to places?

Year 3: Places are both similar and different

Inquiry questions

- What are the main natural and human features of Australia?

- How and why are places similar and different?

- What would it be like to live in a neighbouring country?

Year 4: The Earth's environment sustains all life

Inquiry questions

- How does the environment support the lives of people and other living things?

- How do different views about the environment influence approaches to sustainability?

- How can people use environments more sustainably?

Year 5: Factors that shape the human and environmental characteristics of places

Inquiry questions

- How do people and environments influence one another?

- How do people influence the human characteristics of places and the management of spaces within them?

- How can the impact of bushfires or floods on people and places be reduced?

Year 6: A diverse and connected world

Inquiry questions

- How do places, people and cultures differ across the world?

- What are Australia's global connections between people and places?

- How do people's connections to places affect their perception of them?

  (ACARA 2020a)

Questions

1. What aspects of geography do you think are emphasised in the primary years?

2. Are there areas of geography that you think are missing or underplayed?

Unpacking the core knowledge and suitable case studies for Years 7 to 10

This section sets out an interpretation of the topics for each year level. It begins with an overview of the aims and intentions of the topic and a description of its place in geography. There is commentary on the interpretation of the curriculum and the methods that could be used to develop the core knowledge through case studies.

Links between the content and the concepts inherent to the unit are described. For every topic, the content descriptors (ACARA 2020b) are shown in bold type, followed by our description of the core knowledge and understanding that should be developed in that content descriptor. This is followed by a selection of appropriate inquiry and skills which fit with the content descriptor.

Case studies that are suitable for developing the core knowledge are then suggested. The descriptions of the case studies are aimed at showing teachers some of the ways in which the core knowledge can be developed using geographical thinking and methodology. In some cases, reference to a website or other resource is given. In other cases, we have described a possible case study that could be developed in different ways due to a school's location.

Year 7

The key inquiry questions for Year 7 are:

- How does people's reliance on places and environments influence their perception of them?

- What effect does the uneven distribution of resources and services have on the lives of people?

- What approaches can be used to improve the availability of resources and access to services?

Unit 1 Water in the World

This unit focuses on a number of issues related to the geographical location of water resources and the uses of this essential resource. An inquiry into aspects of this important theme in geography involves the use of many skills, such as local data collecting, interpreting diagrams of water cycle processes and representing information on maps.

The influence of water on human existence and activities is so important that it is a logical topic to cover with Year 7 students. The knowledge and understanding that they gain from this unit will be built on by subsequent units at higher levels in the manner of the spiral curriculum described by Jerome Bruner (1960) (see Figure 2.1).

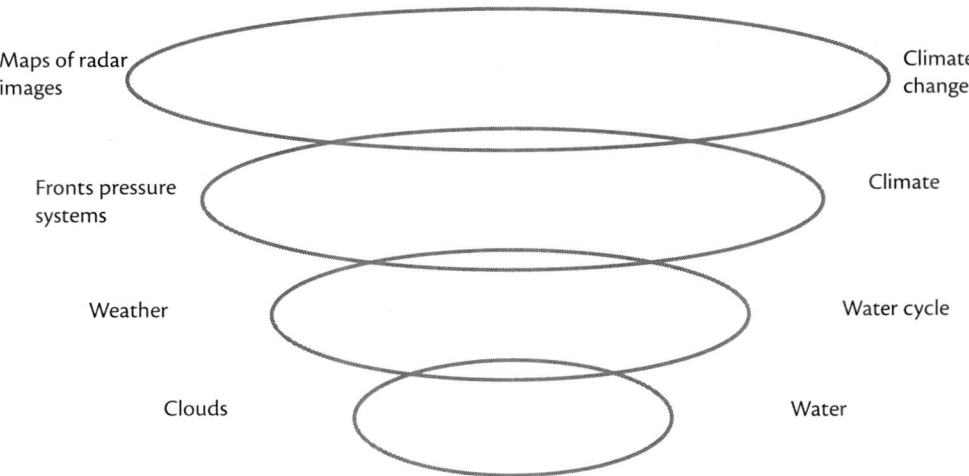

Figure 2.1 The idea of Bruner's spiral curriculum in use

This unit deals particularly with the concepts of space and environment. The importance of space is shown in the maps and diagrams, which show the spatial patterns of water scarcity. Water is such an important feature of environment that this concept is central in the content and case studies discussed below.

Geographical knowledge and understanding

Content descriptor: Classification of environmental resources and the forms that water takes as a resource (ACHGK037)

Students should develop an understanding of the importance of water as a key resource to humans in many ways. They also need an awareness of the many forms of water as well as the changes of form within the water cycle. Case studies of water resources in the local area are a suitable way of developing these concepts while at the same time developing skills of fieldwork observation as part of an inquiry into features of the local area.

Core knowledge:

- Understanding the idea of what a resource is
- Classifying renewable and non-renewable resources
- Surface water, ground water and soil moisture in the water-cycle at a variety of scales

Inquiry and skills:

- Collecting information on surface water, ground water and soil moisture

Case studies:

- **Local area fieldwork observation**: Students with the aid of satellite images should identify examples of surface water. They could then use a base map of the local area and mark on it any rivers and other surface water. With the help of the teacher, they should identify any presence of ground water in the local area, such as springs or lakes. Then, in any local area, they could investigate the presence of soil moisture by digging a few centimetres into the soil. The amount of soil moisture could be described qualitatively in words (wet, moist, dry, parched) or quantitatively using a soil moisture-measuring device available from garden shops. The data from this fieldwork can then be added to the map of surface water and ground water. Patterns and the reasons for them could then be described.

- **Different forms of water throughout the world**: Use a set of maps and diagrams on water in the world to instigate inquiries into how much of the world's water is fresh, what other forms water is found in, which continents have the most and the least water, how much water is used on each continent and the differences in rainfall and evaporation.

Content descriptor: The way that flow of water connects places as it moves through the environment and the way this affects places (ACHGK038)

Students should develop an understanding of the concept of a river basin and learn the major features of it. They should learn to recognise river basins and water features on maps, and appreciate how much our urbanised society alters the flow of rivers for many different purposes. Good examples of these alterations can be found near all of the capital cities and large towns. Aerial photographs and satellite images show the locations of reservoirs in relation to rivers.

Core knowledge:

- River basins, tributaries, estuaries, deltas
- Natural flows and diversions by reservoirs and pipelines

Inquiry and skills:

- Interpret information from weather maps and rainfall radar
- Draw sketch maps of features of river basins
- Draw and interpret graphs of weather statistics

Case studies:

- **Water supplies to a local large city**: For this case study, the teacher would need some information from the state or local water supply authority. In particular, maps of reservoirs, storage tanks, catchment areas and pipelines would be very useful. A base map from Google Maps is a starting point on which the location of all the features mentioned above can be plotted.

- **Fieldwork in local area**: Students observe signage naming rivers, catchments, reservoirs, pipelines, stormwater flows. They could mark the observations of all students in the class onto a large wall map, or a shared digital map (see Figure 2.2). Photographs of such signs could be added to the correct locations.

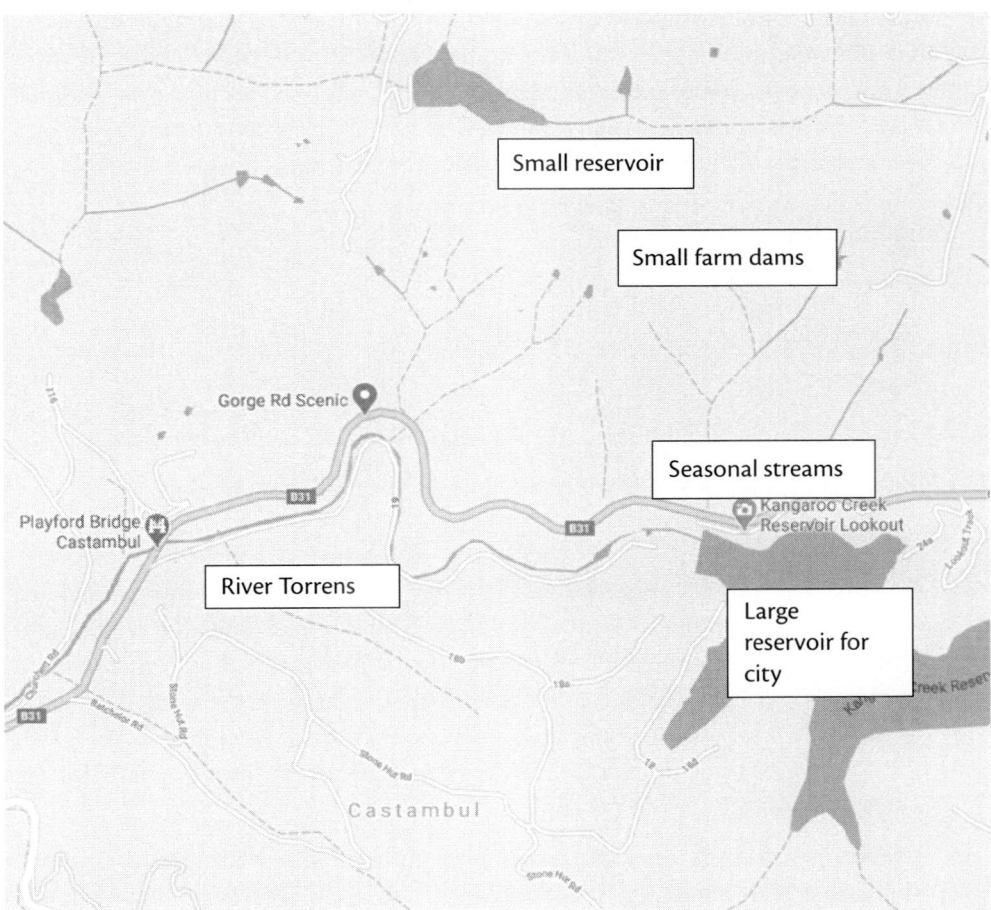

Figure 2.2 Sample of a map with water features named

- **Weather maps, satellite images and rainfall radar from the Bureau of Meteorology:** The water on which we depend for life, food, industry and health ultimately comes from the sky in the form of rain. Therefore, an understanding of causes and patterns of rainfall is clearly linked to a study of water in the environment. The Bureau of Meteorology website shows daily maps and statistics of all places in Australia. Students should become familiar with looking at the local rainfall radar and matching it with their own observations. They should also access the weather maps for the next four days, note how they show changes over the four days, and then make predictions based on the weather maps.

Content descriptor: The quantity and variability of Australia's water resources compared with other continents (ACHGK039)

Rainfall is a central element of weather and climate. An understanding of the causes of rainfall links the water cycle with topography, and illustrates the variability of water supply from place to place and from time to time. Daily weather patterns, seasonal differences, spatial patterns and changes in climate are all illustrated within any inquiry into the water cycle. Contrasting Australia with other continents helps consolidate an appreciation of key issues for Australia.

Core knowledge:

- Causes and seasonal patterns of rainfall
- Rainfall and river patterns in Australia and other continents

Inquiry and skills:

- Drawing diagrams to illustrate the causes of rainfall in contrasting places
- Interpreting statistics of climate and river flows on other continents

Case studies:

- **Contrasts of rainfall, evaporation, rivers and lakes between southern, northern and central Australia:** Students should look at a set of maps of rainfall and evaporation in Australia. They should convert the mapped information into written descriptions of the differences in patterns between the north and south of Australia. From these descriptions, they should think further in an inquiry into the reasons for these differences, and the consequences of the differences for the landscape and the uses of the land.

- **Clouds and rainfall:** An understanding of the formation of clouds and rainfall provides a strong basis for an appreciation of the differences in climate across Australia, and the various effects of climate change. A set of diagrams of cloud and rainfall formation by convection currents, frontal systems and mountain ranges (see Figure 2.3) can be used in explanation and reinterpreted by students. This understanding of processes then logically progresses to comparisons of regions of Australia and contrasting overseas regions.

Content descriptor: The nature of water scarcity and ways of overcoming it, including studies drawn from Australia and West Asia and/or North Africa (ACHGK040)

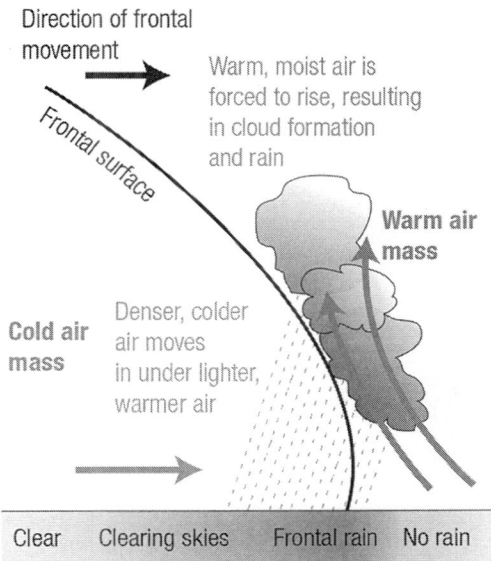

Figure 2.3 The influences of a cold front on the weather

The problem of water scarcity is an important global issue for the present and future. Students need to appreciate the small percentage of the world's water that is fresh drinking water. Political, economic and technological responses to this scarcity illustrate how humans interact with resources and physical geography.

Core knowledge:

- Proportions of fresh water, sea water, ice caps, groundwater on Earth

- Physical reasons and human reasons for water shortages

- Methods of conserving fresh water and using it wisely

Inquiry and skills:

- Analysing use of water at home, at school and in the local area, and proposing improvements

Case studies:

- **Water scarcity in areas of Africa and South-West Asia – conflict and conservation**: Maps of rainfall and evaporation could be analysed to identify the areas of Africa and South-West Asia which have potential water shortage. This inquiry can then be followed by a search for dams, pipelines and irrigation channels, and details of some of the main uses of water for crops and settlements.

- **Water use in industry/Our water footprint**: The amount of water that is used in many industries to produce goods that we use is not often recognised, but it is essential to our way of life. Making clothing, steel, cement and paper requires large amounts of water. More obviously, water is essential to the growth and processing of most of our food. Students can find out figures of how many litres are used in the manufacturing, transport and sale of common objects such as a t-shirt, hamburger or car. The map of virtual water flows around the world introduces and demonstrates the idea of the water footprint for each country.

- **Access to clean water – diseases and hygiene:** Clean water is one of the basic needs of humans. A survey of how much clean water is used in the students' houses provides a starting point for a broader study of parts of the world where clean water is scarce.

Content descriptor: Economic, cultural, spiritual and aesthetic value of water for people, including Aboriginal and Torres Strait Islander peoples and peoples of the Asia region (ACHGK041)

The three curriculum priorities of Aboriginal and Torres Strait Islander peoples' perspectives, interactions with Asia and the importance of future sustainability are all focused on in these studies. Contrasts in the value placed on water, and the techniques of using it and conserving it develop students' complexity of understanding different cultures' use of resources. Irrigation for rice growing is central to the staple food of so many of the population in Asia, whereas the use of irrigation to produce some food crops in Australia is a debated issue.

Core knowledge:

- Dependence on irrigation for food supplies
- Aboriginal and Torres Strait Islander peoples' understanding of water sources
- Conflicting views on uses of rivers, lakes, coastal waters

Inquiry and skills:

- Analysing the reasons for contrasting views of water use and conservation

Case studies:

- **Rice growing with irrigation systems in South-East Asia:** Growing rice using irrigation is one of the most widespread agricultural systems in the world, and is particularly central to the people of South-East Asia. Students should look at maps showing its distribution pattern, photographs and video clips of the landscapes, and graphs of rainfall patterns. The influence of equatorial rainfall, monsoonal seasons and consistently high temperatures are the basic physical influences, and elaborate irrigation systems with terracing and water control provide high yields.

- **Irrigation systems in the Murray-Darling Basin – contrasts with South-East Asia:** Irrigation systems in the Murray-Darling Basin are large in scale and rely on drawing large quantities from the rivers. Students should look for contrasts in scale, climate, intensity of agriculture, type of crop and infrastructure between irrigation in Australia and that in South-East Asia.

- **Aboriginal peoples' knowledge and use of water resources.** It is difficult to generalise about Aboriginal knowledge of water resources because the regions of Australia differ so much in their climate and availability of water. However, it is known that in desert areas, Aboriginal communities survived by having a detailed knowledge of where water might be found in dry seasons, and how it must be conserved and not wasted. The low population density of the dry regions and the ability of these communities to move when necessary also contributed to survival in dry areas.

Content descriptor: Causes, impacts and responses to an atmospheric or hydrological hazard (ACHGK042)

Learning to understand the causes and effects of an atmospheric or hydrological hazard is an effective way of appreciating the interactions of physical and human factors within the environment. In almost all examples of disasters resulting from such hazards, human factors such as settlement location and protection, building laws, society responses and defences, weather observations and communication are all just as significant as the physical causes of the hazard.

Core knowledge:

- Physical causes and spatial patterns of either droughts, storms, tropical cyclones or floods
- Contributions of humans to the risk factors of a particular atmospheric hazard

Inquiry and skills:

- Collecting and presenting data on a hazard in a variety of forms
- Interpreting the links between location, settlement and climate change on a hazard

Case studies:

- **Examples from Australia and overseas of whichever hazard has been chosen:** Maps of the location patterns of either droughts, storms, tropical cyclones, or floods are important resources for this, as are descriptions and measurements of the seriousness of the events. Human contributions to these hazards should be examined and assessed in relation to the physical factors.

- **Drought in Australia and Africa:** The contrasts between the effects of droughts in Australia and Africa focus on the level of technology used in food production, and the ability of farmers to withstand droughts through whatever resources are available to them. Statistics of changes of production levels, availability of drought relief and the role of governments and NGOs should be studied.

- **Path and effects of a recent tropical cyclone:** A case study of a recent tropical cyclone in northern Australia should use weather maps, wind and rainfall statistics, and photographs. The paths of tropical cyclones can be analysed, together with their causes. Inquiries into the effects on cities and settlements in the north of Australia should produce evidence of problems of siting of buildings and the need for more building regulations, as well as inhabitants' need for up-to-date information on cyclone movements.

Pause and think

1. In one or more of the sections of Unit 1 described above, choose a case study and suggest how you could link the development of both knowledge and skills.

2. Choose one of the sections in Unit 1 and, after thinking about the core knowledge, the inquiry and skills, and the case studies, plan how you would create a series of lesson plans to teach this section of the unit.

3. Fieldwork is suggested in a few sections of Unit 1. Discuss the possibilities for such fieldwork in your local area.

4. There are a number of different types of maps mentioned in this unit. What other kinds of maps and map skills could be used in teaching this unit?

Unit 2 Place and Liveability

This unit develops the central concept of place in geography by linking it to liveability. Each place in the world has a unique identity or character of its own. It also has an attraction and level of liveability that is very different for different groups of people. This unit examines some aspects of liveability, such as accessibility to desired services, environmental quality and social connectedness.

Inquiry into the characteristics of liveability in the local area forms a sound base for looking at liveability elsewhere in the world. Critical use of the international surveys of liveability is included here, and inquiries that focus on different perceptions of place by different groups of people are also important.

Geographical knowledge and understanding

Content descriptor: Factors that influence the decisions people make about where to live and their perceptions of the liveability of places (ACHGK043)

The concept of place is developed well by focusing on the perceptions of places held by different groups of people. This kind of inquiry highlights the characteristics of the places where people live, and illustrates the different perceptions of these places by different groups of people.

Core knowledge:

- Different meanings of liveability for adolescents, young families, retirees, migrants, refugees, wealthy, poor, different cultures

- Different ways by which liveability can be measured

Inquiry and skills:

- Collecting and recording relevant geographical data

- Presenting ideas in a range of communication forms

Case studies:

- **Different perceptions of liveability in your local area:** A starting point in inquiry could be to have students list the factors they see as making their local area desirable for them (as teenagers) to live in, and those that make it undesirable. They should then make the same two lists from the point of view of an older person, a recently arrived migrant and an unemployed person. The lists can be compared and evaluated to identify similarities and differences.

- **Characteristics of place:** Students can be asked to choose a place that they know well or in which they have a particular interest. To convey a clear description of the features of the place, they could create a brochure, a set of images, a video, an interview or a slide presentation. The place might be their local area, or a far-off place where they would like to live.

Content descriptor: The influence of accessibility to services and facilities on the liveability of places (ACHGK044)

A key factor of liveability is accessibility to services, work, family and friends. This accessibility is highly dependent on access to transport – both private and public. Other factors include the age group and income group of particular people. This inquiry into accessibility in the local area and in a remote location with a very unusual transport system reveals the effect of community decision-making.

Core knowledge:

- Services and facilities that increase liveability

- The influence of transport patterns on accessibility

Inquiry and skills:

- Representing spatial data by constructing appropriate maps

- Collecting and recording geographical data from primary and secondary sources

Case studies:

- **Services and transport patterns in the local area:** On a base map of the local area, students can draw the routes of public transport, and mark the busiest roads and large parking areas. They can then locate 'dead areas', which are remote from public transport. On another map of the same scale, services such as shops, medical facilities, schools and parks should be marked. Areas that are remote from these can be identified. This leads to reflection on the effects of the identified differences on liveability.

Content descriptor: The influence of environmental quality on the liveability of places (ACHGK045)

Most people rate environmental quality high in their choice of residence, close behind location of work and affordability. People with higher incomes are more likely to rate it more highly because of the choices provided by wealth. There are many factors that raise or lower the environmental quality of places.

Core knowledge:

- Understanding of common types of pollution – air, water, soil, noise, visual

- The link between pollution and health

Inquiry and skills:

- Collecting and recording geographical data from primary and secondary sources

Case studies:

- **Environmental quality:** Assessing factors such as air pollution, tree shade, noise pollution, shelter from rain and wind, smells in the school grounds or the local area is a practical fieldwork activity that heightens student sensitivity to factors of environmental quality. This can be followed by inquiry into the environmental quality of other places in the world by using maps and pollution data.

- **Collecting comparative data in fieldwork:** A survey of liveability in the local area, or another comparative area, could be done using the survey sheet shown in Figure 2.4. The recording and processing of information should lead to an evaluation of the measures suggested in this survey, and the proposal of other measures that could be used.

Criteria	Score				
	1 Very poor				
	2 Poor				
	3 Satisfactory				
	4 Good				
	5 Very good				
Category 1: Environment					
• Temperature/humidity	1	2	3	4	5
• Quality of urban design and architecture	1	2	3	4	5
• Access to parks and gardens	1	2	3	4	5
• Amenity of streetscapes	1	2	3	4	5
• Maintenance of public spaces	1	2	3	4	5
Category 2: Cultural					
• Quality of community recreational facilities	1	2	3	4	5
• Availability of places of worship	1	2	3	4	5
• Diversity and quality of restaurants	1	2	3	4	5
• Availability of public libraries	1	2	3	4	5
• Range of entertainment venues	1	2	3	4	5
Category 3: Economic					
• Range of employment opportunities	1	2	3	4	5
• Access to affordable housing	1	2	3	4	5
• Access to consumer goods and services	1	2	3	4	5
Category 4: Infrastructure					
• Quality of road network	1	2	3	4	5
• Quality of public transport	1	2	3	4	5
• Quality of telecommunications infrastructure	1	2	3	4	5
• Availability of good quality housing	1	2	3	4	5
• Provision of utilities – water, electricity, sewerage	1	2	3	4	5
• Availability of cycle ways	1	2	3	4	5
Category 5: Education					
• Availability of private schools	1	2	3	4	5
• Quality of public schools	1	2	3	4	5
• Access to post-school educational institutions	1	2	3	4	5
Category 6: Healthcare					
• Quality of private healthcare	1	2	3	4	5
• Quality of public healthcare	1	2	3	4	5
• Availability of aged-care facilities	1	2	3	4	5
Category 7: Law and order					
• Amount of petty crime	1	2	3	4	5
• Amount of violent crime	1	2	3	4	5
• Graffiti and vandalism	1	2	3	4	5
• Sense of personal safety	1	2	3	4	5

Figure 2.4 A sample survey sheet to gather data on factors relating to liveability

Content descriptor: The influence of social connectedness and community identity on the liveability of place (ACHGK046)

Most people prefer to live in places that have good accessibility to friends, family and people of similar aspirations to themselves. Of course, this is not always possible, and it might be thwarted by work locations and income. Maps of spatial patterns of residents show clear groupings of people of similar incomes, similar cultural backgrounds and similar age cohorts. They also indicate how people are limited in their residential choices by the cost of housing and rents.

Core knowledge:

- Places that are exclusive or inclusive to different people
- Social isolation and inclusiveness in society

Inquiry and skills:

- Collecting appropriate information from primary sources
- Interpreting geographical data to identify spatial distributions

Case studies:

- **Patterns of social groups in Australian cities**: There is plenty of mapped data of populations in the Social Atlases of Australia, which have been compiled for Australian cities and local government areas. Students can look for patterns of age groups, rent costs, migration from other countries and incomes. Hypotheses about the reasons for spatial correlations between some of these factors can be suggested and investigated.

- **Measures of liveability in large cities throughout the world:** There are many lists available that rank cities in terms of liveability. These are often compiled from the point of view of a high-income/highly mobile businessperson who is interested in living in another city, but they can provide valuable information on which factors have been used in their compilation, such as environment, education facilities, healthcare and infrastructure. Students should locate the cities on a world map, noting their location in terms of climate, coasts, proximity to other cities, and the country in which they are situated.

- **Investigation of the importance of connectedness and community identity:** Students can be asked to reflect on the period when the COVID-19 pandemic led to a lockdown in their local area and much of the world. They could ask their parents to remind them of the effects it had on both good and bad connections with other people. Questions about the breaking of connections (such as group gatherings) and making of new connections (possibly online) might suggest some of the ways in which social connections interact with liveability.

Content descriptor: Strategies used to enhance the liveability of places, especially for young people, including examples from Australia and Europe (ACHGK047)

The issue of enhancing the liveability of places is of central importance to town planners, local councils and developers who want to sell real estate. People at different stages of their life want different kinds of housing, and different facilities. Income levels and expectations have major influences on perceptions of liveability.

Core knowledge:

- Environmental sustainability and liveability
- High-, low- and medium-density housing
- Facilities for different groups of people

Inquiry and skills:

- Presenting arguments and ideas to suit a particular audience or purpose

Case studies:

- **Facilities for different groups of people in a city – young, old, disabled, poor, etc.:** Schools, childcare centres, aged-care homes, playgrounds, leisure centres, sports grounds, community centres, services for the disabled – each of these is necessary for different people. What are the most desirable facilities for teenagers? Are they provided in the local area? What happens if they are not available?

- **High-, low- and medium-density housing:** In some local areas, students will be very familiar with the full range of densities of housing, and the change from low density to higher density with urban renewal, gentrification and urban consolidation. In other areas, photographs and brochures of developments might be needed to widen appreciation of options in housing. The advantages and disadvantages of each density of housing can be listed and debated.

Short-answer questions

1. In one of the topics of Unit 2 described above, choose a case study and suggest how you could develop a concept that was inherent to the core knowledge of that section.

2. Choose one of the topics in Unit 2 and, after thinking about the core knowledge, the inquiry and skills, and the case studies, plan how you would create a series of lesson plans to teach this section of the unit.

3. Using information about the local area is suggested in a few sections of Unit 2. Discuss the possibilities for such ideas in your local area.

4. Digital maps are used in a number of sections in this unit. What aspects of digital maps can best show city features?

Integrating Geographical Inquiry and Skills with Knowledge and Understanding

Consult the Australian Curriculum: Geography website for the Inquiry and Skills appropriate for Year 7.

Pause and think

Read through the extract from the Australian Curriculum: Geography. Make a list of the 'inquiry action words' suggested in Inquiry and Skills. Then make another list of the 'skills resource words' (such as 'maps', statistics'). How can these two lists help give you some understanding of how to use the statement on inquiry and skills in classroom lesson planning?

Short-answer questions

1. Suggest methods that could be used in a Year 7 class to show understanding of the geographical processes influencing a place.

2. Suggest a technique involving maps that could demonstrate a student's skills of analysis of the features of a river basin.

3. Suggest an issue related to one of the two Year 7 units that would show the effects of social, economic and environmental influences.

Year 8

The key inquiry questions for Year 8 are:

- How do environmental and human processes affect the characteristics of places and environments?

- How do the interconnections between places, people and environments affect the lives of people?

- What are the consequences of changes to places and environments and how can these changes be managed?

Unit 1 Landforms and Landscapes

This unit provides an introduction to geomorphology, enabling learning about processes that shape landforms and landscapes. The landscapes and landforms of places on Earth give each place a unique character, and at the same time give us clues about the Earth's processes, which have produced these landscapes. In some places, it is easy to see these processes in action as earthquakes, volcanoes, landslides and coastal change.

This unit particularly expands understanding of the concepts of place, environment and change. The unique characteristics of landforms and landscape of each place are discussed. These landforms are a central part of the environment, and are always in a state of change – sometimes invisibly slow and sometimes uncontrollably fast.

The significance and management of landscapes is also taught in this unit, because each unique place is often developed by a combination of natural forces and human management.

Skills of using simple contour maps, drawing and interpreting diagrams of landforms, and using large data/real-time maps of hazard events can be developed within this unit.

Geographical knowledge and understanding

Content descriptor: Different types of landscapes and their distinctive landform features (ACHGK048)

Pre-existing knowledge of the appearance of landscapes such as coasts, deserts, mountains, plains and river valleys can be expanded and related to particular places. Landforms that are part of each of these landscapes can be described, sketched, modelled or mapped.

Core knowledge:

- Differences between landscapes in different places
- Names and descriptions of characteristic landforms, such as dunes, cliffs, volcanoes, canyons, valleys, gorges, waterfalls, flood plains and billabongs

Inquiry and skills:

- Sketching, mapping and describing photographs of landforms

Case studies:

- **Landforms as part of landscapes:** Figure 2.5 shows combinations of landforms that are characteristic of certain landscapes. Students could find photographs of real examples of the landforms named on the diagrams and make a note of their location. Drawing sketches of examples of these landforms helps in the recognition of them and in the understanding of how they relate to each other in a landscape.

Content descriptor: Spiritual, aesthetic and cultural value of landscapes and landforms for people, including Aboriginal and Torres Strait Islander peoples (ACHGK049)

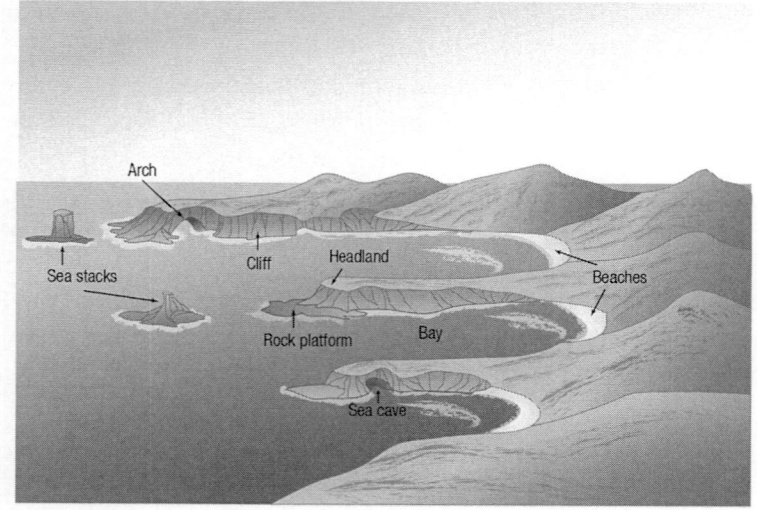

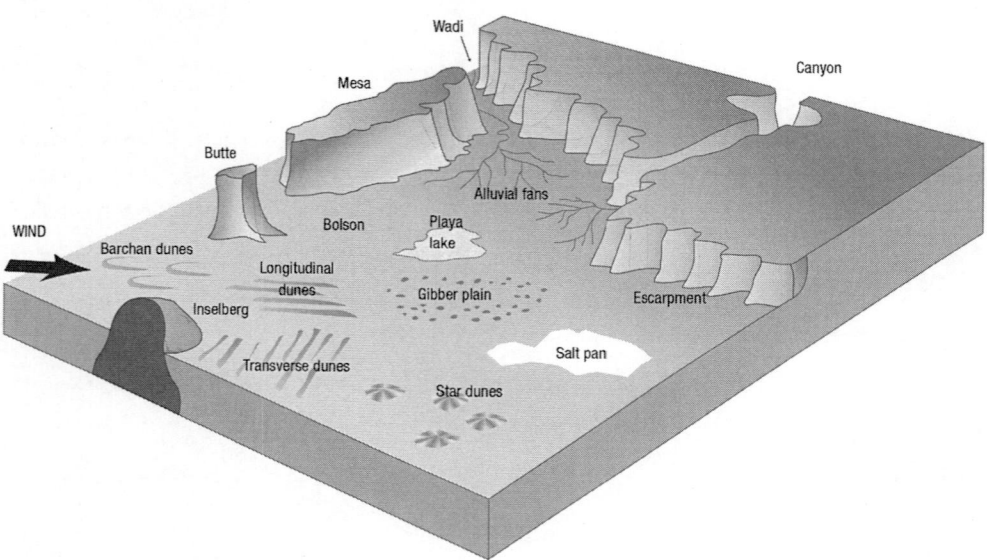

Figure 2.5 Typical landform of a coastal area (top) and an arid area (bottom)

The landscape that surrounds us or is part of our nation or heritage is very important to most people. National anthems of many countries contain specific references to their landscapes, and the literature and art convey the images of those landscapes in different ways. Even though most of us live in cities, we retain an image of the natural landscape as defining our nation. Aboriginal and Torres Strait Islander peoples have a very strong relationship with Country, the land of their group and ancestors.

Core knowledge:

- Links between specific landforms/landscapes and cultural values
- Understanding of the importance of certain landforms/landscapes to particular cultures

Inquiry and skills:

- Collecting relevant geographical information from primary and secondary sources

Case studies:

- **Iconic landforms in stories, art, and popular culture:** Photographs of easily recognised landforms/landscapes such as Uluru, the Grand Canyon, Mount Everest, Bondi Beach, Victoria Falls, the Great Barrier Reef, Monument Valley, Rotorua, Kakadu, Iguazu Falls, the tepuis of Venezuela, Yellowstone, or the white cliffs of Dover can provide a stimulus to thinking about their significance to people living near their location, and people throughout the world.

- **Links between cultures and landscapes:** Landscapes of particular characters are often an important part of the setting of well-known books and films. Examples of these can be used to demonstrate the ways in which landscapes and cultures are linked. The Australian outback, the American West, the Canadian arctic, Siberia, the African savanna, the Amazon River, the Nile River and the English Lake District are the settings for stories and legends of various cultures.

- **Country of Aboriginal and Torres Strait Islander peoples:** Country is a very important value for Aboriginal and Torres Strait Islander peoples, and it needs to be discussed and understood. The example of contrasts in the meaning of Uluru to Aboriginal and Torres Strait Islander peoples and non-Indigenous people can be examined in relation to the ban on climbing.

Content descriptor: Geomorphic processes that produce landforms, including a case study of at least one landform (ACHGK050)

Visualisation and understanding of the geomorphic processes that have contributed to the shape of landforms makes the appearance of landscapes much more interesting. An understanding of how rocks are formed, changed and broken down provides a sound basis for recognising landforms and interpreting the landscape.

Core knowledge:

- Simple understanding of processes of rock formation – tectonic, sedimentary, metamorphic
- Simple understanding of processes of weathering, erosion and deposition
- Linking of these processes in the formation of a particular landform

Inquiry and skills:

- Using maps and diagrams to explain geomorphic processes

Case studies:

- **What processes caused a beach/island/mountain range/waterfall/sand dune to be where it is?** This could be an effective inquiry question, which could lead to a consideration of erosion processes such as running water, wave action and wind, rock building from volcanoes, faulting, folding, sedimentary layers, heat and pressure.

- **Case study of a landform:** Uluru, the Grand Canyon, the Matterhorn, Wilpena Pound, the buttes of Monument Valley, the tepuis of Venezuela, Mount Vesuvius, the White Cliffs of Dover – any well-known landform can be studied and explained in terms of processes.

Content descriptor: Human causes and effects of landscape degradation (ACHGK051)

Landscapes are changed by humans, particularly in areas of high population density. Building, farming, mining, recreation, transport and changes to plants and animals all have potential effects on landscapes. They may cause changes to erosion, silting, runoff, river flows and habitats.

Core knowledge:

- Built environments create hard surfaces, which change water runoff.

- Changes in water runoff may cause more erosion, more silting and changes to river patterns.

- Desertification is a widespread type of degradation of semi-desert caused by a combination of damaging treatment of the land and changes to the climate.

Inquiry and skills

- Interpreting geographical data to propose explanations for spatial distributions and trends

Case studies:

- **Desertification:** Semi-desert regions have borderline rainfall. Climate change has altered the rainfall, and unsuitable uses of the land have changed many of these areas into deserts. An inquiry into the patterns of rainfall in North Africa shows examples of these changes. Maps of the changes in rainfall isohyets over the last 40 years are a valuable starting resource.

- **Changes to runoff patterns:** People who live in a city are often oblivious to the changes that the hard surfaces such as bitumen roads, paved paths and dense building patterns have made to water runoff. In a rural area, light rain seeps straight into the soil and heavy rain runs across the surface to lower ground and eventually either finds its way into a river or soaks into the soil on the way. There is little unpaved area in densely populated cities, so that rainfall has to be got rid of by large-scale storm water drainage systems of underground pipes and artificial surface channels. Thus an artificial drainage system replaces a natural system.

- **Changes to landforms:** Rivers and coasts in their natural state are continually changing their shape because of changes in the amount of rainfall and erosion and deposition from day to day, season to season and year to year. These natural changes are often forgotten when people build on riverbanks, coastal cliffs and sand dunes. The more defensive structures that are built to protect houses and roads, the more effects the rivers and coastal forces have on the landscape.

Content descriptor: Ways of protecting significant landscapes (ACHGK052)

There are certain environments that are highly valued because of their beauty or uniqueness or recreational characteristics. However, they are often threatened by human activities, ignorance or greed. Ways of protecting such environments have therefore been devised, including declaration of national parks, world heritage status and laws on land use.

Core knowledge:

- There are widely different views about the value of certain landscapes.
- Protection of valued environments is necessary but complex.

Inquiry and skills:

- Proposing actions to protect significant landscapes

Case studies:

- **An inquiry into the protection of one highly valued landscape:** A case study of an environment such as Kakadu, Uluru, Tasmania's wilderness, the Flinders Ranges, the Kimberley, the Blue Mountains or the River Murray could highlight the differences in viewpoints, and the policies that have been used to protect that environment.

Content descriptor: Causes, impacts and responses to a geomorphological hazard (ACHGK053)

Geomorphological hazards include earthquakes, volcanoes, landslides and tsunamis. Each of these is a natural feature of the physical landscape and cannot be prevented by humans. However, their effects can differ greatly due to the activities of humans. Settlement patterns, poverty, infrastructure and the level of resources have major influences on the severity of these natural events.

Core knowledge:

- Movements and forces in the Earth's crust cause natural occurrences of earthquakes, volcanic eruptions, tsunamis and landslides. These events cannot be prevented by human actions, but their effects can be modified.

- Humans can alter the impacts of these hazardous events by improving planning, building types and regulations, emergency responses and community education.

Inquiry and skills:

- Drawing and interpreting block diagrams

- Inquiring into hazard spatial patterns with digital maps

Case studies:

- **Earthquake and tsunami in Japan, 2011:** The links between the causes of an earthquake and a tsunami are shown in this inquiry. It also clearly demonstrates the damage done to highly populated and low-lying areas. The location of a nuclear power plant in the area destroyed by the tsunami adds to the understanding of the interaction between physical and human factors in geomorphological hazards.

- **Location patterns of volcanic eruptions and earthquakes:** World maps (digital or paper) of the locations of earthquakes and volcanoes are ideal resources for an inquiry into the links between these patterns (see Figure 2.6). They illustrate

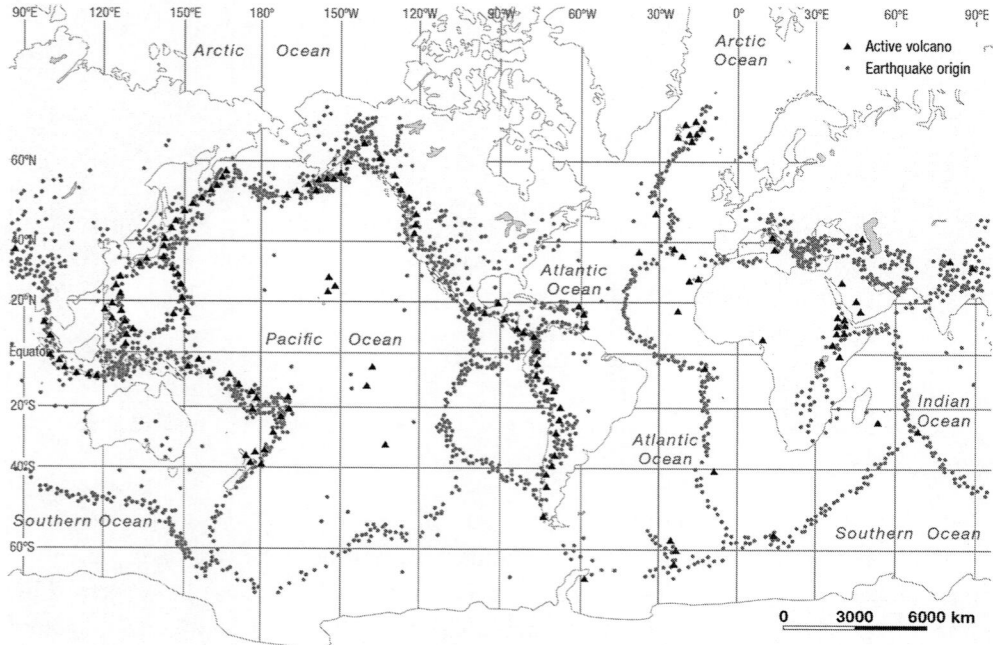

Figure 2.6 Global pattern of earthquakes and volcanoes

the correlation between earthquake and volcanic hazards along the edges of the Earth's tectonic plates. Block diagrams of the forces acting along the plate boundaries can be used to explain the movements of earthquakes and the landforms caused by volcanoes.

Short-answer questions

1. Make a list of landforms that you think Year 8 students could sketch and label. Suggest a plan for a lesson that would expand the students' knowledge of names and shapes of landforms.

2. Contour maps show vividly the shapes of landforms such as volcanoes, cliffs, canyons and rivers. Plan a learning activity that uses contour maps as a key resource in learning about landforms.

3. Discuss and suggest some examples for debate topics that relate to the damage that can happen to famous landforms from an overload of tourists.

4. Choose two videos on YouTube of geomorphological hazard events that you consider show either the causes or the effects of the event very clearly. Compare your choices with those of others.

Unit 2 Changing Nations

Since 2010, more than 50 per cent of the world's population has lived in cities. Urbanisation is continuing at a fast rate – by 2030, that figure will be 60 per cent. But this process of urbanisation has great variations across the world. This unit examines the process of urbanisation across the world, and specifically in Australia, the United States and China. The differences and similarities between patterns of population growth and change within these countries lead to an understanding of interconnections between rural and urban populations and economies. The concept of change is central in this unit.

Geographical knowledge and understanding

Content descriptor: Causes and consequences of urbanisation drawing on a study from Indonesia or country of the Asia region (ACHGK054)

Urbanisation is the growth of cities and the shift of people from rural areas to cities. There are factors pushing people away from their traditional rural home regions and factors pulling people towards cities. Many of these push and pull factors relate to jobs and income.

Core knowledge:

- Urbanisation is a continuing process that is increasing the percentage of the world's population who live in cities

- Urbanisation occurs at different rates in different countries and at different times

- Push and pull factors both have influences on people's change of residence

Inquiry and skills:

- Interpreting geographical data using qualitative and quantitative methods

Case studies:

- **Reasons why so many people migrate to cities:** An understanding of the push and pull factors that are powerful influences on people in rural areas can be illustrated with an example of the fast growth of Jakarta, by migrants moving in from the rural areas of Indonesia.

- **Urbanisation and the environment:** Large cities create their own environments in terms of microclimate, water runoff, plants, landforms, animals and birds. Their growth involves the taking of land that might previously have been farmland, natural vegetation or wetland, and converting it to a built-up area. Before and after photographs of large cities, or the local area, can give clear impressions of these kinds of changes.

- **Urbanisation:** The process of urbanisation has produced a world where the majority (over 50 per cent) of people live in cities. Most population growth now occurs in cities of developing countries. Maps and data can be used to find the countries with extremes of urbanisation, and to identify the largest cities within them.

Content descriptor: Differences in urban concentration and urban settlement patterns between Australia and the United States of America, and their causes and consequences (ACHGK055)

Australia and the United States are of a similar size in area, but have very different population patterns. The reasons for this are found in physical landscape, occurrence of resources and settlement history.

Core knowledge:

- The spatial patterns of population in Australia and the United States

- The influence of physical factors and historical factors on settlement patterns

Inquiry and skills:

- Interpreting statistical maps

Case studies:

- **Inquiry into the difference in population patterns between Australia and the United States:** Maps of population density and location of large cities are good starting points for an inquiry into the differences between Australia and the United States. These can be complemented with maps of physical landscape, resources, transport patterns and annual rainfall. The exploration can continue with dates of the founding of what have become the major cities in each country.

- **Cities of Australia and the United States:** Through popular culture, the names of many cities of the United States are known to students, but this knowledge is likely to be superficial. A deeper knowledge can be developed by looking at the location, population, state and Google Street photographs of these cities, and comparing them with an Australian city with one similar characteristic (climate, population size, function and jobs, state capital).

Content descriptor: Reasons for, and effects of, internal migration in both Australia and China (ACHGK056)

Most countries have distinct patterns of movement of people from one region to another, and from rural to urban areas, or vice versa. One of the most important causes of such internal migration is the growth of new jobs in some areas, and the loss of jobs in others. At a stage in life when employment becomes less important, retirement or lifestyle might be the reason for migration to highly desired locations.

Core knowledge:

- Patterns and causes of internal migration

Inquiry and skills:

- Identifying and proposing explanations for spatial distributions, patterns and trends

Case studies:

- **The growth of the urbanised area of Hong Kong/Shenzen in southern China:** Hong Kong was a British colony until 1997, when it reverted to become part of China. Hong Kong has always been an important trading port, as well as an industrial city. Over the last 50 years, it has been joined by an urban cluster of cities in China's Guangdong Province, across the border from Hong Kong, producing goods for the world market, such as toys.

- **Movements of Aboriginal and Torres Strait Islander peoples:** Those who identify as Aboriginal or Torres Strait Islander peoples have migration patterns that contain factors additional to those of non-Indigenous people. These factors include the emotional attachment to the Country or Place of their family and ancestors, and the dispossession from their Country or Place by farmers, pastoralists and miners. These are in addition to the need for employment, opportunities for their children and availability of housing.

- **Movements for employment in Australia:** Contrasts between permanent migration versus fly-in/fly-out (FIFO) work patterns can lead to an understanding of alternatives of residence for workers in remote areas. The increase of working at home using technology (stimulated by the COVID-19 pandemic) is also altering the links between work and residence for those in many other types of employment.

> **Content descriptor:** Reasons for, and effects of, international migration into Australia (ACHGK058)

There are many different reasons for people immigrating to Australia, and a listing of these (permanent, work-related, student, family reunion, refugees), together with statistics of arrivals, helps an inquiry into immigration. The attraction of large cities for migrants is an important fact in the growth and diversity of these cities.

Core knowledge:

- Reasons for international migration to Australia
- Effects of international migration to Australia

Inquiry and skills:

- Collecting and recording relevant data
- Applying geographical concepts to draw conclusions based on the analysis of data

Case studies:

- **Family migration to Australia:** Australia has had many waves of immigration from many different countries at different periods. The first migrants from a region are usually followed by successive groups of family members. The importance of family migration to Australia can be noted by analysing the statistics of place of birth of Australian residents and place of birth of their parents.

- **Visual evidence of effects of international migration:** In many parts of Australian cities and towns, there is plenty of evidence of the influence that immigration has had on Australian way of life. Examples include foods, shops, places of worship, sports, newspapers and television newscasts. A listing of these from students' observations in the local area would be a suitable field study inquiry in many locations.

Content descriptor: Management and planning of Australia's urban future (ACHGK059)

Australia's large cities are spread mostly far apart along the coast. They are mainly state capitals, with all the economic advantages this entails in attracting population. Plentiful supply of land has led to urban sprawl. All cities have tackled these problems by planning for more urban infilling, better transport systems, and more high- and medium-density housing. They have also attempted to make cities 'greener', both in vegetation and in water and energy use.

Core knowledge:

- The meaning, cause and effects of urban sprawl

- Current planning and large-scale schemes used in Australian cities

- Making cities more sustainable

Inquiry and skills:

- Interpreting aerial photographs and maps of urban patterns

Case studies:

- **High-, low- and medium-density housing:** Photographs of examples of each of these different densities of housing lead to the question of personal preferences versus community advantages. The environmental effects of each of these can be examined.

- **Sustainability in cities:** Ways of making cities greener, by using renewable energy, lowering water use, planting more trees and initiating better public transport, are among the schemes to increase sustainability that can be investigated. Examples of these can be found in most of the state capital cities.

- **Urban sprawl:** Most large cities have areas of urban sprawl, where suburban growth has occurred without control. Students can look at areas of urban sprawl using aerial photographs on Google Earth. They can look for size of blocks of land, type of housing, location of shopping centres and other services, and areas of parks and recreation. Discussion of transport systems, infrastructure such as water, gas, electricity, sewerage, internet access and typical demographics of the inhabitants can suggest other characteristics that the aerial photographs do not reveal.

Pause and think

1. What differences would you want to make to this unit if you were teaching it in a rural school compared with an urban school?

2. Which large cities have you personally visited or lived in? What are some particular characteristics of these cities that you could use as vivid examples that would connect with students?

Short-answer questions

1. Suggest methods that could be used in a Year 8 class to show understanding of the ways in which physical processes of erosion and deposition produce landforms of particular shapes.

2. Suggest a technique involving graphs of various kinds that could illustrate a student's understanding of the process of urbanisation and the growth of megacities.

3. Suggest an issue related to one of the two Year 8 units that would show the effects of social, economic and environmental influences.

Year 9

The key inquiry questions for Year 9 are:

- What are the causes and consequences of change in places and environments and how can this change be managed?

- What are the future implications of changes to places and environments?

- Why are interconnections and interdependencies important for the future of places and environments?

Unit 1 Biomes and Food Security

Food security is one of the most important things upon which we all rely, but it is also one of those most often forgotten. This unit lets students inquire into the degree of security of food supplies, both for Australia and for other parts of the world. It first establishes the links between different biomes and the foods that can be grown in them. With the speed of population growth since the Industrial Revolution, there was pressure to grow more food, and this has led to significant alterations of biomes and sophisticated use of technology to increase food production. This unit examines the concepts of environment, change and sustainability in the inquiry into the future of food supplies.

Geographical knowledge and understanding

> **Content descriptor**: Distribution and characteristics of biomes as regions with distinctive climates, soils, vegetation and productivity (ACHGK060)

Biomes are regions characterised by particular climate, soils and vegetation. All of these are linked as elements in the broad ecosystem of the region. A knowledge of the spatial patterns of biomes across Australia and the world is the basis for understanding how they are used for food production now and in the future.

Core knowledge:

- Biomes and ecosystems as organising concepts
- The broad spatial pattern of biomes in Australia and the world

Inquiry and skills:

- Interpreting patterns and correlations on maps
- Applying geographical concepts to draw conclusions based on the analysis of data

Case studies:

- **Spatial patterns of biomes:** An inquiry into spatial correlations between climate and latitude, followed by spatial correlations between climate and vegetation, leads to an understanding of the broad patterns of biomes. This knowledge can be deepened by looking at climate graphs and vegetation types for a pair of places remote from each other but located in similar biome regions.

- **Flow of energy in ecosystems:** Ecosystems are powered by the sun, which provides light, warmth and power for the water cycle. The flow of energy from the sun through the plants, into the soil and into animals is an important concept for an appreciation of the stability and delicacy of each ecosystem.

- **Productivity of biomes:** Because of the differences in sunlight, temperatures and moisture, there are great contrasts between the productivity of different biomes. Regions that have high temperatures and rainfall have the greatest productivity. Students could make hypotheses about the ranking of biomes in order of their productivity, then check their hypothesis against the data shown in Figure 2.7.

> **Content descriptor**: Human alteration of biomes to produce food, industrial materials and fibres, and the use of systems thinking to analyse the environmental effects of these alterations (ACHGK061)

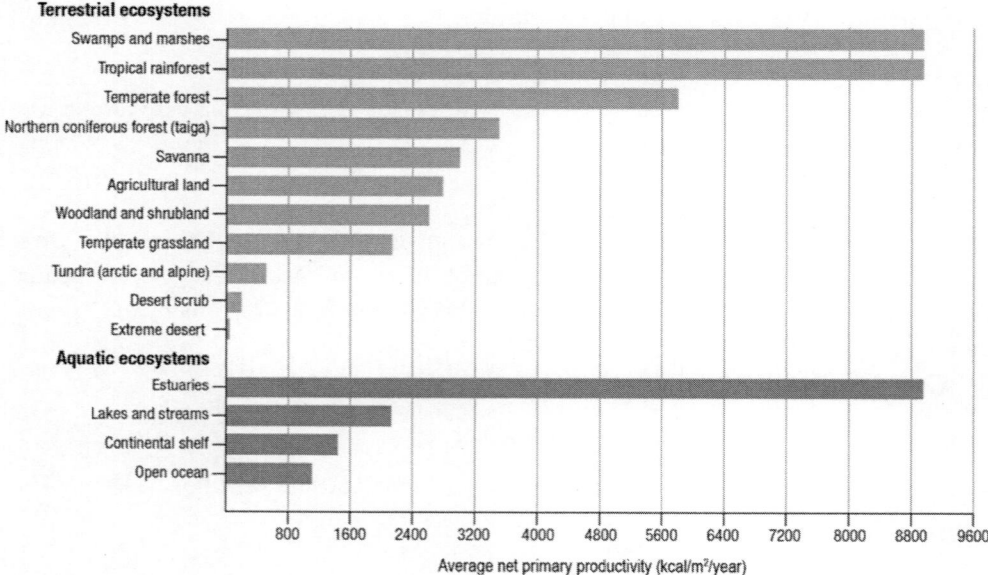

Figure 2.7 Productivity of ecosystems

It is important to develop an understanding that the food we eat is mostly produced in distinctive biomes, from which it enters a processing chain and a supply chain to reach consumers. The production of food in large quantities has a significant effect on the biome.

Core knowledge:

- The effects of large-scale food production on biomes

- Interdependence in food supply chains

Inquiry and skills:

- Using flow diagrams to represent data and processes

- Studying detailed maps and aerial photographs for evidence of changes

Case studies:

- **Alterations of biomes by humans:** Food growing on a large scale is necessary because of the growth of the Earth's population. Low-density populations can produce food without much change to the natural plants, soils and animals, but the need to produce food for millions of urban dwellers leads to large-scale clearance of natural vegetation, drastic changes to the soil, damming rivers for irrigation and using large quantities of chemicals. An intensive map study of an area of large-scale agriculture reveals evidence of many of these changes.

- **Effects of intensive rice growing on the biomes of South-East Asia:** In much of South-East and Eastern Asia, rice is the staple cereal. This creates a high demand for the growing of large amounts of rice and its availability throughout the year. The climate of high temperatures and high rainfall is suitable, and rice can be grown intensively. However, it is best grown on flat ground that can be flooded or drained, and where all other vegetation is usually cleared. This means that in hilly areas the slopes are often terraced and set up with irrigation systems. Many of these altered landscapes have been functioning for many years.

Content descriptor: Environmental, economic and technological factors that influence crop yields in Australia and across the world (ACHGK062)

The natural environment in which crops grown for food or fibre are planted is only one of the factors influencing the yield of the crop. Farmers are always trying to obtain higher yields with the use of new technology, better use of resources, new varieties of seed, types of fertiliser and soil improvement, as well as the application of new research. However, the influence of each of these factors on yield has varied from place to place and from time to time. The economic influences on crop growing are primarily the price that farmers can get for their crop, and the costs to the farmers of all the inputs, and each of these factors can vary greatly over a short time.

Core knowledge:

- The production of crops for food and fibre is controlled by a balance of environmental, economic and technological factors

Inquiry and skills:

- Interpreting geographical data to propose explanations for patterns and trends

Case studies:

- **Technology and crop yields:** From the time just before the Industrial Revolution, new technology has played a major role in changing farming and crop yields. New varieties of plants, new ways to plant seed and harvest crops, new ways to plough the ground, new fertilisers, new ways of irrigation and the substitution of machines for manual labour are among the changes brought about by technology. However, there have been downsides to each of these as well. Students could draw up lists of advantages and drawbacks of these new methods of farming.

Content descriptor: Challenges to food production, including land and water degradation, shortage of fresh water, competing land uses and climate change, for Australia and other areas of the world (ACHGK063)

Most of our food is now produced by large-scale farming. This agriculture has a high capital input (complex machinery) and a low labour input (machines instead of people). Farming has extended into large areas of marginal land once all the best land is in use.

This has led to problems with soil quality, regular water supply, and conflict with other uses such as mining and housing. The effects of climate change on food production are being monitored. For some regions, this will mean great changes.

Core knowledge:

- The challenges posed by the large-scale farming which produces most of our food

- Interactions between producing food and its effects on land degradation

- Climate changes that will lead to major changes in food production

Inquiry and skills:

- Proposing action in response to a contemporary geographical challenge

- Drawing conclusions based on the analysis of data

Case studies:

- **The effects of climate change on food production:** Climate change affects rainfall as well as temperatures, and there have been many studies of the likely effects of climate change on specific locations in Australia and across the world. The belts where crops such as wheat, corn, cotton and fruits have been grown will move with climate change. Students can postulate the likely impacts this will have on farmers, service towns, transport systems, exports and costs for consumers.

- **Land degradation:** Over-use and poor management of the land lead to degradation (downgrading) of the quality of the soil. Examples of land degradation include soil erosion, acidification of soil, soil salinity, lack of organic humus. Dryland salinity in parts of South Australia and Western Australia has been caused by the clearance of too many trees with deep roots that soaked up ground water. Without these trees, the groundwater and the salt within it rises to near the surface of the soil and makes the soil too saline to grow crops. Students should describe and explain the processes using diagrams such as Figure 1.6 in Chapter 1, then find diagrams of irrigation salinity and explain the difference.

Content descriptor: The capacity of the world's environments to sustainably feed the projected future global population (ACHGK064)

The population of the world is predicted to grow from 7.7 billion in 2020 to 8.5 billion in 2030 (United Nations 2019). Although the growth rate of population has been slowing down in recent years, there will still be nearly an extra billion people to be fed in the next 10 years. This can happen through increased yields, clearing more land for farming, using more chemicals and fertilisers, producing different foods, using genetically modified (GM) crops or using technologies and methods that have not yet been fully developed. Each of these methods has environmental effects, which need to be assessed before the method is used widely.

Core knowledge:

- An understanding of the graphs and statistics of population growth

- Appreciation of the benefits and risks of various methods of increasing food supply

Inquiry and skills:

- Taking account of environmental, economic and social considerations, predicting the expected outcomes of proposals

- Developing significant questions and planning an inquiry based on them

Case studies:

- **Can northern Australia produce more food?** The area of Australia north of the Tropic of Capricorn has high temperatures and high rainfall, factors which lead to fast and luxuriant plant growth. However, it is a region located a long distance from the large cities in southern Australia. For a hundred years, it has been regarded as a region that should be able to produce more food, and there have been many projects and schemes. An inquiry into schemes such as the Ord River Scheme, Humpty Doo rice growing, sugar cane in North Queensland and the introduction of Zebu cattle can reveal both successes and failures, and the factors causing these.

- **Should we be using a greater variety of food sources?** This interesting question can suggest a lot of investigation into the addition of insects and different plants into our diets. Food has a strong cultural dimension in terms of what people prefer to eat, but this can be expanded by new experiences, acceptance of other cultures and the relative cost of different foods. An inquiry, discussion and debate on this question should suggest new thinking to students.

Pause and think

1. Students who have always lived in the city sometimes have little experience of the sources of their food. What teaching in this unit could you use to widen their understanding and appreciation of where their food comes from and how it gets to them?

2. The COVID-19 pandemic highlighted the importance of supply chains that get food and necessary provisions from sources to processors, wholesalers and retailers. Australia is in the fortunate position of producing much more food than it consumes. However, there were shortages of some goods. In what ways could you involve students in thinking about the effects of future disruptions of supply chains of food?

3. Old geography courses and school textbooks over 50 years ago used to require students to name the agricultural and industrial products of many countries. There are both educational and pragmatic reasons why this is no longer done. What do you think are some of these reasons?

Short-answer questions

1. In what ways can the effects of climate change be integrated into this unit?

2. What is meant by the 'productivity of ecosystems' and the 'flow of energy'? What methods would you suggest to help students understand these concepts?

3. What overall conclusions can be derived from the graph of productivity (Figure 2.7)?

Unit 2 Interconnections

This unit focuses on one of the key concepts of geography: interconnection. Geography deals in a holistic way with many different interconnections between people, places and processes. This unit deals with some of the major interconnections between people and places, in particular the movements of people, goods, information and ideas across the world. The introductory content looks at the concept of place, and the ways in which our perceptions of place differ from person to person. Interconnections are examined on scales ranging from local to global.

Geographical knowledge and understanding

Content descriptor: The perceptions people have of place, and how these influence their connections to different places (ACHGK065)

We all have our personal geography – our own knowledge of places and our own amount of interaction with a range of places. Whenever we hear what other people think about places, we realise that perceptions of the same place range widely from person to person. Perceptions of your local area almost certainly vary enormously because of the experiences, the age, the wealth, the family background, the beliefs and many other characteristics of different people. In the same way, perceptions of far-off places differ depending on accuracy of knowledge, personal contact, family history, personal interests and many other factors. It is important to recognise the variations in perceptions.

Core knowledge:

- How perception of places differs greatly from person to person
- Methods of building accurate knowledge of other places

Inquiry and skills:

- Representing spatial distributions by constructing maps at different scales
- Identifying and proposing explanations for spatial patterns and inferring relationships

Case studies:

- **Perceptions of the place where you live:** An inquiry into perception and inter-connections in the local area can be done by devising a set of simple questions. These could include asking family and friends to comment on characteristics of the local area in terms of appearance, liveability, services, shops, schools and transport, among others. The answers can be classified by age, gender and any special needs of respondents. Patterns of similar perceptions by people of similar background should be looked for.

- **Perceptions versus accurate knowledge about other places:** Mention the name of a place, like New York, Paris, Mumbai, Hawaii, Cairo or Shanghai, and most people have an image or a few facts in their mind. However, the image and facts are really perceptions and may be far from the truth. These perceptions have a large influence on people's attitude to other places. They influence their desire to visit the place, buy goods from the place, read about the place or express opinions about it. The use of non-stereotypes is essential in order to challenge inaccurate perceptions.

- **Mental maps:** Everyone has mental maps of places in which they live, and of the rest of the world as they know it. Mental maps are not accurate maps; they are maps of perceptions. Ask students to sketch their mental maps of the local area, their region, their state, Australia, or the world. The sketches show a wide variety of perceptions and knowledge. The reasons for the individual differences may be particular interests, family connections, life experiences or any combination of similar factors. Students may recognise from this exercise how prejudiced our mental maps often are.

- **Moving within your personal space:** How much do we actually know about the area in which we live? How much do we physically move about within the town/region/city where we live? These are interesting questions with which to challenge students. Using a Google map as a base map of the local area, each student can draw lines and shading to indicate areas in which they regularly or occasionally move, and areas that they know well, or that they know very little about. An extension of this exercise is drawing similar maps to show the places that are further away with which students have connections.

Content descriptor: The way transportation and information and communication technologies are used to connect people to services, information and people in other places (ACHGK066)

The increase in globalisation over the last few decades has increased interconnections between many people in many countries. International travel has become cheaper, faster and more common. Buying goods from other countries (components, wholesale, retail) has increased interconnections. The internet has made it easy and cheap to make contact with relatives, friends, business associates, corporations, accommodation providers and sales outlets anywhere in the world.

Core knowledge:

- Global patterns of interconnections through transport

- Interconnections across the world through information and communication technology

Inquiry and skills:

- Using real-time big data apps to access information and recognise spatial patterns

- Using community information to reflect on current trends in the use of technology and its effects

Case studies:

- **Patterns of air travel using flight tracking apps:** Any of the apps that show all aircraft in the sky in real time can produce the data for this case study. The broad patterns of the number of aircraft on the main routes in Australia, the United States and Europe can be examined. The routes and stopping places for long-haul places can be identified and explained. The areas without many aircraft in their skies raise interesting questions. From these investigations, generalisations about the places with very high or very low levels of interconnections can be made.

- **Changes in workplace caused by the COVID-19 pandemic:** Because most nations adopted a policy of lockdown and social distancing in 2020 that continued in 2021, working from home using personal computers and the internet became more common, and accelerated the adoption of this practice by many businesses. The change needed high-volume and fast internet connections throughout the world and larger amounts of cloud storage. Students can discuss the disruptions to traditional ways of working and communicating that have been experienced recently, and predict some of the likely effects on place of residence, types of employment and interconnections with others.

Interconnections with other places and people through trade have been important throughout history, and have increased rapidly during recent globalisation. The reasons and effects of trade in goods and services are an important piece of knowledge necessary for future decision-making.

Core knowledge:

- Changes in patterns of goods and services – local and global
- The location of places that export most consumer goods and places that import most consumer goods

Inquiry and skills:

- Interpreting cartograms
- Interpreting quantitative flow maps

Case studies:

- **Interpreting cartograms to show interconnections:** The website WorldMapper (worldmapper.org) draws cartograms of the world's nations showing the relative size of their characteristics. Use the website to create cartograms of the countries that produce a particular good, and of those of the nations that consume the same good. Compare the patterns.

- **Interpreting information using quantitative flow maps:** The flow of goods, services and people between countries and regions is often shown effectively by bars, which are scaled to represent the quantity of people or goods. Atlases and websites can supply many of these, clearly showing the interconnections – or the lack of them – between countries.

China is a major political and economic power in the world, and has many interconnections with Australia, including imports and exports. China has been a major trading partner with Australia, but there are changes occurring with regard to this that need to be noted. The fast growth of China's economy is having major global effects, which change from year to year.

Core knowledge:

- Australia's interconnections with China

- The growth of China as a major centre for manufacturing and trade

Inquiry and skills:

- Interpreting and analysing geographical information to make generalisations and inferences

Case studies:

- **China's Belt and Road Initiative:** China has put into operation a very large regional plan of funding ports, roads, bridges, shipping and other transport to connect Asia more easily with Africa and Europe. The plan includes six major corridors to improve trade. Such a scheme will change the pattern of world trade and bring greater connectivity to many countries. This is a suitable case study for inquiry into interconnections between nations, and the benefits and problems can be drawn out of the information available.

Content descriptor: The effects of people's travel, recreational, cultural or leisure choices on places, and the implications for the future of these places (ACHGK069)

Global tourism has been growing at a rapid rate (apart from the period of COVID-19), as illustrated in Figure 2.8. Tourists have brought prosperity and much foreign cash to the 'honey pot' locations, but have also caused many environmental and social problems to these locations. Large cruise ships, luxury hotels, adventure trips to remote locations, increases in international flights and changes to traditional cultures are just some of the effects of increases in global tourism.

Core knowledge:

- The changing pattern of world tourism

- The effects of tourism on places and the environment

Inquiry and skills:

- Presenting findings, arguments and ideas in a range of communication forms

- Interpreting geographical data and proposing explanations for patterns and trends

Case studies:

- **World tourism – its changing pattern and importance:** World tourism has grown steadily over the last 50 years, with only two blips on the graph – one in 2008 caused by the global financial crisis and a much greater fall in 2020 caused

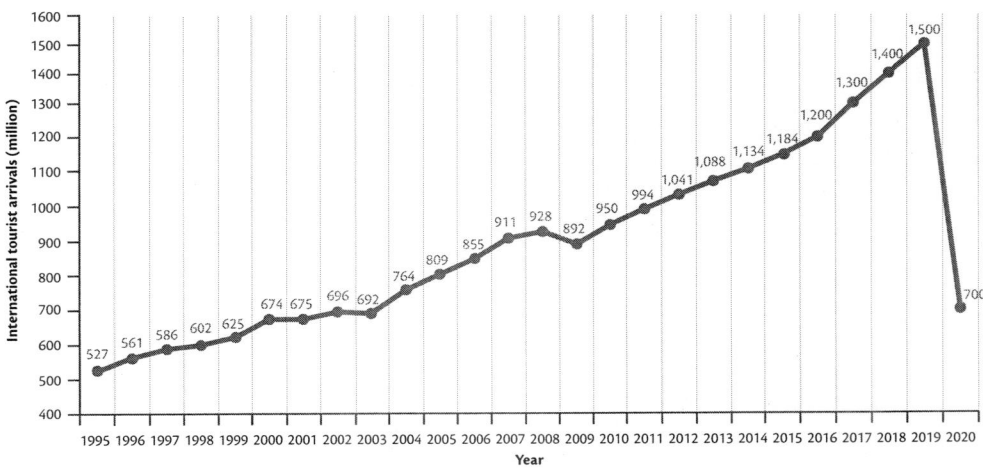

Figure 2.8 Growth in world inbound tourism

by the COVID-19 pandemic. Tourism has become a major source of income for many countries and regions. Figure 2.8 shows the growth of tourism from 1995 to 2020. Analysis of these figures can produce inquiry questions about the causes of changes, the growth of the cruise ship industry, the countries providing the most tourists and the countries attracting the most tourists.

• **The benefits and damages of tourism – Venice:** Most people in the world have seen pictures of Venice in Italy and have heard of its canals, gondolas and beautiful old buildings. It is high on the list of must-see tourist destinations. Thousands of people arrive every day to see the sights. But Venice is on an island with one road to the rest of Italy; most of its streets are tiny lanes intended only for pedestrians; its canals are also mostly narrow; and most of its hotels are small. All of these characteristics add to its charm and its attractiveness for tourists, but the sheer volume of tourists overwhelms them. When cruise ships flood the city with 3000 tourists and tower over the beautiful old buildings, the problems caused by mass tourism are not hard to see. Students could inquire further into the issues of Venice as a tourist 'honey pot', including the physical damage done to buildings by the combination of sinking, and annual floods and high tides. They could also compare other places that have similar problems.

Pause and think

1. What do you understand by the term 'personal geographies'? If you were to draw your mental map of the spaces in which you move, how extensive would it be now compared with five or 10 years ago?

2. Which are the parts of the world about which you have a reasonably accurate perception? Does this come from personal visits, family background, reading, films and TV, contact with friends, internet connections or other interconnections? Your students' interconnections with other parts of the world will vary greatly between schools and locations, and may need sensitive handling.

3. There are a number of different kinds of interconnections mentioned in this unit. List some of these different kinds of interconnections. Which would be familiar to students? Which would more likely be unfamiliar?

4. With which parts of the world would you predict that the students you are teaching have connections? With which other regions or countries might you expect them to be familiar?

Short-answer questions

1. What methods would you suggest would be effective in helping students understand how to interpret cartograms?

2. What are some other advantages and disadvantages, not mentioned above, of tourism to developing countries?

Year 10

The key inquiry questions for Year 10 are:

- How can the spatial variation between places and changes in environments be explained?

- What management options exist for sustaining human and natural systems into the future?

- How do world-views influence decisions on how to manage environmental and social change?

Unit 1 Environmental change and management

This unit builds on previous studies of environment and change to examine in greater detail one particular environment, in order to inquire into the changes to that environmental type and the ways in which people can manage that environment to maintain its sustainability. The unit first looks at environmental changes and stability in general, then explores the contrasting views of different people, including the environmental management practised by Aboriginal and Torres Strait Islander peoples.

To engage students in concentrated inquiry and to encourage deep knowledge, there is a choice of one out of five environments to be used as a detailed case study. The choice of environment is with the teacher, taking into account location of the school, the background knowledge of the students and the availability of resources. The chosen environment is examined for its systems, management processes and the success of the management.

Geographical knowledge and understanding

Content descriptor: Human-induced environmental changes that challenge sustainability (ACHGK070)

This unit focuses on sustainability, and the environmental changes that threaten sustainability, so this introductory content is important to help students understand these broad concepts. Students who have studied previous units from the Australian Curriculum: Geography should have a general understanding of the concept of sustainability, but this knowledge can be deepened by linking biomes, ecosystems and biodiversity. The concept can be developed further by treatment of the interaction between ecological, economic and social sustainability.

Core knowledge:

- Sustainability – its meaning and its complexity
- Effects of human-induced environmental changes
- The importance of biodiversity

Inquiry and skills:

- Developing significant questions that apply geographical methods and concepts
- Reflecting on and evaluating the findings of an inquiry

Case studies:

- **Searching for sustainability:** An inquiry into what is meant by sustainability uncovers both knowledge and lack of it. Although the focus of this unit is on ecological sustainability, it is important for students to understand the importance of economic sustainability and the interrelationship this has with ecological sustainability. In a similar way, the links between both of these and social sustainability should be examined. A key principle of sustainability is the maintenance of biodiversity. Each plant and animal occupies a niche in its ecosystem, supplying food, shelter and support to other plants and animals. When a species is wiped out, the links in the ecosystem are broken and the whole system becomes more frail. Examples of loss of species such as the Dodo, Tasmanian Tiger, Lesser

Bilby and Japanese Sea Lion could be used as an introduction to an inquiry into other losses of biodiversity.

- **How humans create environmental changes – pollution and degradation:** Until the Industrial Revolution and the Medical Revolution that followed closely in its wake, people tended to accept pollution as a natural part of existence. The high death rates and prevalence of illness, epidemics and dirty water supplies were part of everyday life. With the Industrial Revolution came dense settlement of some areas, and extreme levels of pollution of the water, air and soil. The environment of many areas was changed greatly by mining, industry, land clearance and population growth. In agricultural areas, food production had to be increased to feed the extra people, and this led to significant changes to the land. Examples of pollution and degradation of the environment that can illustrate this include open-cut coal mining in Victoria, air pollution in Beijing, pollution by industry in California and radioactive material in the sea after the Fukushima tsunami in Japan.

- **Climate change and its effects:** Many students at this year level will have some knowledge of climate change, but because of the community disagreements about causes, effects, seriousness, speed and intervention, students' knowledge will be affected by their beliefs and attitudes. The wise teacher will be aware of the strength of emotions that are apparent concerning this topic. One approach is to present the statistics that are relevant, and invite students to carry out an inquiry of their own into the effects of what is shown by the statistics. Another approach is to focus on the actions that governments in Australia and overseas are taking about climate change, and attempt to match these with possible futures.

Content descriptor: Environmental world-views of people and their implications for environmental management (ACHGK071)

The future of the Earth and its environments will be determined by the beliefs and actions of people. Any discussion of environmental issues reveals the diversity of views that exists about human uses of the Earth's resources. When the human population of the Earth was small, the damage to environments and the management of sustainability were not issues because resources seemed infinite, the new regions that could be settled and developed seemed endless and new technologies were constantly being devised. Now it is realised that many resources are not infinite, and there are few new regions that can be settled. New technologies are still being developed, but their emphasis is changing towards using technology to produce sustainability of resources and environments rather than trashing them in a frantic race for resources. Some appreciation of the reasons for the contrast in world-views is necessary.

Core knowledge:

- Environmental management – aims and techniques

- Why are there great contrasts in world-views of sustainability and environmental management?

Inquiry and skills:

- Developing geographical questions and planning an inquiry

- Reflecting on and evaluating the findings of an inquiry

Case studies:

- **How can humans better manage environments?** An understanding of environmental management can be demonstrated with an inquiry-based case study of a small area. The key natural and human features should be documented and the pressures on them identified. The links between all of these can then be noted, as can any obvious effects. This inquiry activity leads to its logical conclusion of suggestions for improvement in management based on the data gathered. Such management might include limiting access to an area, banning certain uses of the area, revegetating, altering water flows, changing road systems, increasing maintenance and changing laws. High-level decisions may have to be made by governments; specialist decisions should be based on detailed research.

- **Is sustainability possible?** An inquiry into sustainability – economic, environmental and social – should produce large amounts of information about different attitudes towards sustainability practices. Students can be encouraged to evaluate the information they find about sustainability to provide their answer to this key question. The deep and powerful knowledge that students have developed from all their previous geographical studies should provide a framework for this.

> **Content descriptor**: The Aboriginal and Torres Strait Islander peoples' approaches to custodial responsibility and environmental management in different regions of Australia (ACHGK072)

Archaeological evidence shows that Aboriginal and Torres Strait Islander peoples have lived in Australia for close to 80 000 years, in a wide variety of environments. During that time, they have passed on to each generation an understanding of their Country, and an appreciation of the techniques needed to survive in it.

Core knowledge:

- How did Aboriginal and Torres Strait Islander peoples manage the environments of Australia in almost 80 000 years of occupation?

Inquiry and skills:

- Collecting and recording relevant geographical information

- Presenting findings and explanations

Case studies:

- **Management of the environment by a particular group of Aboriginal and/or Torres Strait Islander peoples:** A local group or one from an environment of particular interest could be chosen. Food, shelter, water supply, movement and knowledge of seasons could be studied. Particular examples could be the knowledge of tropical seasons by the Gagadju people, the knowledge of resources in extreme aridity by the Dieri people and the use of coastal and river environments by the Narrindjeri people.

For the following content descriptions, teachers and students select one of these types of environment for detailed study:

- land environments

- inland water environments

- coastal environments

- marine environments

- urban environments.

Pause and think

1. The Australian Curriculum: Geography gives teachers a choice of one of five different environments to undertake a detailed study. Would you choose the one you know best, the one in which your students are most interested or the one they have not learned much about in their previous study? Explain your choice.

2. What fieldwork activities could be incorporated into one of these detailed studies? What locations outside of Australia could be used as effective contrasting or comparative examples?

Content descriptor: The application of systems thinking to understanding the causes and likely consequences of the environmental change being investigated (ACHGK073)

Whichever environment is chosen, the first task is to develop an understanding of the operation of systems within it, and the human involvement with these systems. This unit could be interwoven with the next two, focusing on the same environment but developing knowledge and ideas about different approaches to it.

Core knowledge:

- **Land:** Changes to the land from clearance, agriculture, fire

- **Inland water:** The significance of catchments in the water cycle system

- **Coast:** Waves and the sand system in the erosion and deposition of coasts

- **Marine:** Reefs, islands, currents and tides in the marine system

- **Urban:** Systems within cities

Inquiry and skills:

- Interpreting geographical data and other information, looking for causes and consequences

Case studies:

- **Land – effects of land clearance:** Thinking about systems such as ecosystems, soil systems and biomes such as forests and deserts can lead to an understanding of the key elements and interactions within these systems. If one or more of the interactions is altered by land clearance, it impacts all other elements and inter-actions. This can be illustrated vividly by system diagrams.

- **Inland water – pollution in catchments:** The catchment areas of rivers are large areas that often are not recognised for their integral role in the purity of fresh water in the river, yet any pollutants that enter the soil or tiny streams may eventually flow into the main river.

- **Coasts – the system of erosion and deposition in coasts:** Coasts get their shape and appearance from a combination of factors, including rock type, rainfall, vegetation, protection or exposure to the wind and waves. The system of wave action causing both erosion and deposition is a key system in coastal geography.

- **Marine – wastes and pollution in the oceans:** The ocean currents are a huge system that shifts water from one ocean to another, having major effects on climate and sea life. They also shift human-made pollution around the world. The patterns of ocean currents should be looked at on a world map or globe so that the size and complexity of the ocean system can be appreciated.

- **Urban – urban systems:** Urban areas are complex systems, containing their own ecosystems (trees, gardens, animals, soils), infrastructure systems (roads, water pipes, sewerage, power lines) and land-use systems (housing, shops, offices, parks, recreation). Most of these are managed and controlled by local laws, but sometimes the interaction of these different systems is ignored in planning.

Content descriptor: The application of geographical concepts and methods to the management of the environmental change being investigated (ACHGK074)

This unit continues the study of the chosen environment, but with an emphasis on management. Examples of management that could be studied are described below.

Core knowledge:

- **Land:** Vegetation management

- **Inland water:** Improving water quality in catchments

- **Coast:** Coastal maintenance

- **Marine:** Laws and restrictions about using the seas

- **Urban:** Urban planning for sustainable living

Inquiry and skills:

- Applying geographical concepts to draw conclusions based on the analysis of information

Case studies:

- **Land – management of vegetation clearance:** Whether the natural vegetation is tropical rainforest, mallee scrub, savannah grassland or any other type, it is the vegetation best suited to that climate. In many biomes, most of the natural vegetation has been cleared for settlement and agriculture, so there is conflict over the management of the remaining vegetation. Studies of the Amazon Basin, the forests of eastern Australia and the eucalypt woodlands of South Australia and Western Australia all provide information about conservation and management.

- **Inland water – improving water quality in catchments:** Students have learned in earlier years about the processes of the water cycle and its importance in our lives. They should have an understanding that there are strong links between the flows of water in rivers, underground, in the soil and in the atmosphere. There has been increasing public recognition that the quality of water in rivers is dependent on what is put into the soil and the atmosphere within the catchment area.

- **Coast – hard and soft engineering solutions to coastal maintenance:** Contrasts between hard and soft engineering used in the management of sandy beaches clearly demonstrate different kinds of management. Hard engineering includes the building of defensive walls to prevent waves from eroding the coast, and building systems of groynes to stop sand from drifting too far. Soft engineering includes replanting vegetation to encourage sand dune growth and preventing sea grass from dying so that the sand cycle is conserved.

- **Marine – conservation of reefs, fishing areas and the oceans:** Until relatively recently, the oceans were regarded as an infinite sink where wastes of every kind could be dumped, and fish could be caught in great quantities. Now the realisation has dawned that this cannot continue. Laws have been framed by nations and by international bodies to control the amount of dumping and the scale of fishing. Many reefs have been affected by coral bleaching, a consequence of climate change. An inquiry into the zoning laws and spatial pattern of zones in the Great Barrier Reef provides an excellent example of an attempt to manage a multi-use ocean resource.

- **Urban – planning for cities:** By their nature, cities take over the land on which they are built, and convert most of it to hard surfaces of buildings and streets. Very little remains of the previous ecosystem, although city dwellers want parks, trees, birds, sunlight, clean air and quiet areas among the urban environment. How this can be achieved within the urban system is a puzzle that confronts planners, officials, politicians and other decision-makers.

> **Content descriptor:** The application of environmental, economic and social criteria in evaluating management responses to the change (ACHGK075)

This unit builds on the previous content by focusing on an evaluation of the management of the particular environment. Questions about past, present and future management by individuals, business firms and governments are raised.

Core knowledge:

- **Land:** National parks and reserves
- **Inland water:** Using and conserving water
- **Coast:** Conflicts between development, conservation, recreation.
- **Marine:** Conflicts between tourism, fishing, conservation
- **Urban:** Sustainable cities; green cities

Inquiry and skills:

- Presenting findings, arguments and ideas using geographical terminology

Case studies:

- **Land – national parks and reserves:** Should national parks be closed to the public and only used to conserve ecosystems? Or should they be open to visitors to enable people to gain an appreciation of the beauty and fascination of the ecosystems and landscapes within them? An inquiry into the different types of national parks and reserves in Australia, the United States, the United Kingdom and countries in Africa raises many questions about their purpose and effectiveness.

- **Inland water – what water uses should get priority?** The most crucial requirement that we have of water is that it be pure enough to drink and in ready supply. Yet the amount of water we use for drinking is only a very small percentage of the water supply we expect to have available for other uses such as industry, agriculture, cleaning, transport, recreation, food production and many other unseen uses. Examples of local rivers (such as Parramatta, Yarra, Swan, Brisbane, Torrens or Derwent) can be studied in inquiries with a mixture of fieldwork and secondary sources. Opinions and facts about which uses should get priority can be an outcome of these inquiries.

- **Coast – conflicts between development, conservation, recreation:** How can the tension between recreation and conservation in coastal regions be resolved? Beaches that are close to cities provide case studies for inquiry and suggestions for management. The suggestions should take into account the environmental systems (beach, waves, dunes, cliffs, plants), the economic criteria (costs, sources of funds, multi-uses) and social criteria (recreational uses, nearby population density, transport systems).

- **Marine – conflicts between tourism, fishing, conservation:** There are many areas of seas and oceans that are highly valued resource locations for contrasting uses. In many areas, commercial fishing has taken many species almost to extinction, and fish farming of various kinds has had to be substituted. Tourists also put pressure on the oceans in the wastes they create and the damage done to heavily visited reefs. An inquiry into the conflict between uses of the seas and oceans can reveal many issues relevant to particular locations.

- **Urban – can cities have greater sustainability?** Cities are the powerhouses of the economies of many countries. Food production is mostly done outside of cities, while their major market is the urban population. People are becoming much more concerned about making cities more environmentally sustainable by planting more trees, retaining more open space, growing gardens on rooftops and using more solar energy panels. Students could suggest other methods or investigate particular examples of new housing or commercial developments.

Short-answer questions

1. Choose one of the five environments specified and suggest how the last three content descriptors can be linked together.

2. What are some vivid examples of good and bad environmental management that can be chosen for students to analyse?

3. Name some of the systems that operate within the environments specified.

Unit 2 Geographies of Human Wellbeing

The concept of human wellbeing is an important part of the teaching of geography because it not only expands students' knowledge of the extremes of the human world, but also deepens their ability to empathise with people whose wellbeing is low, and to examine ways of improving human wellbeing. The differences in human wellbeing across Australia and the world develop the concept of space and spatial variation. The inquiry into reasons for great variations in wellbeing develops the concept of social sustainability.

Geographical knowledge and understanding

Content descriptor: Different ways of measuring and mapping human wellbeing and development, and how these can be applied to measure differences between places (ACHGK076)

To understand differences in human wellbeing, a number of statistical measures are used. Students need to understand how these are derived, and to appreciate their good and bad points. The geographical approach is to display such statistics on a map to help in developing understanding of the spatial patterns of human wellbeing.

Core knowledge:

- Understanding the use of measures such as Human Development Index and GDP per capita
- Familiarity with the UN Millennium Goals

Inquiry and skills:

- Skills of reading statistical data and mapped data
- Skills of interpreting cartograms
- Using websites and apps that present data on human wellbeing in vivid ways

Case studies:

- **Using statistical data in inquiries into human wellbeing:** The maps in Figures 2.9 and 2.10 are examples of the spatial pattern of two frequently used measures of human wellbeing. Together with tables of the most recent statistics of these measures of human wellbeing, they provide plenty of data for inquiry into the differences between continents and regions, and arguments about the validity of various measures. In addition to the maps shown here, statistics of a 'happiness index' are also interesting to examine.

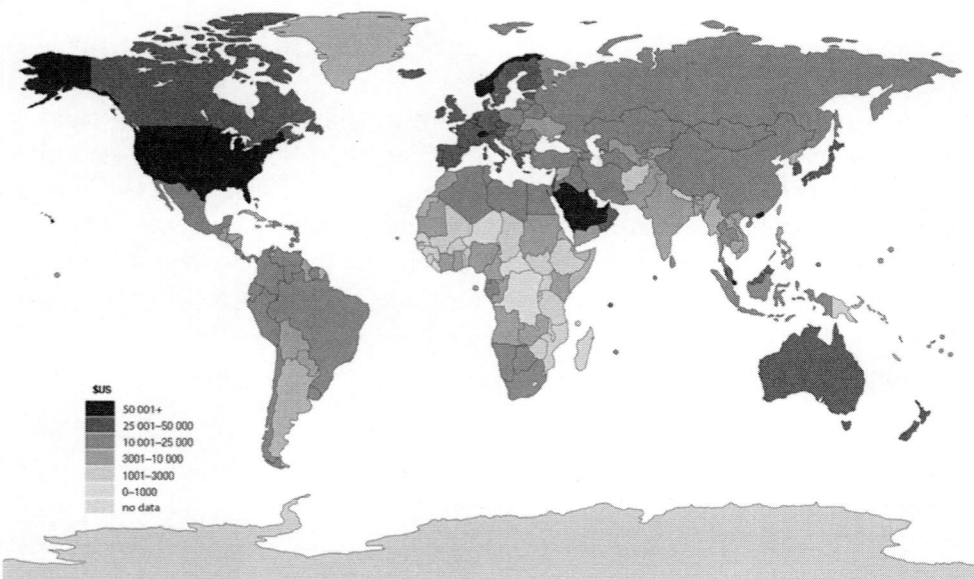

Figure 2.9 Gross national income (GNI) per capita, 2015

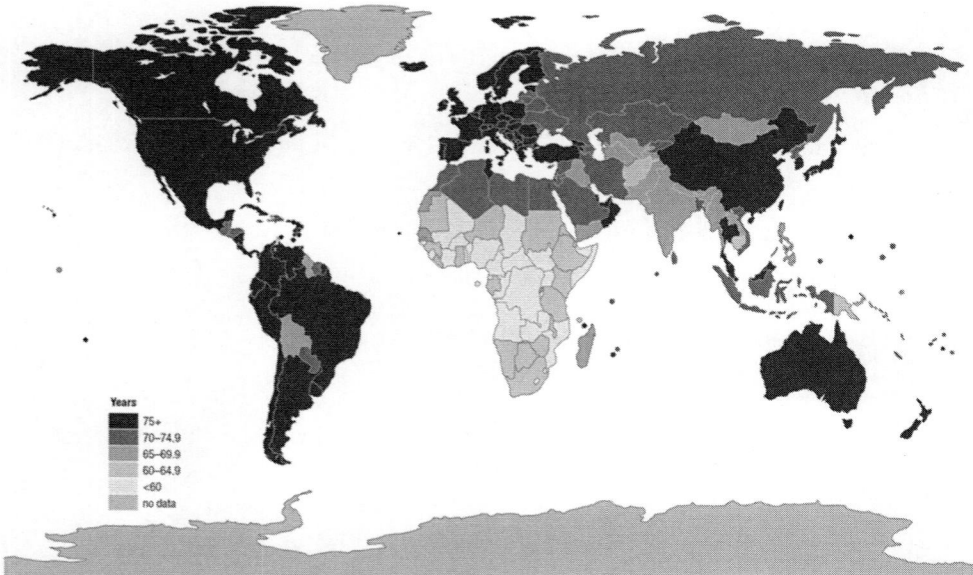

Figure 2.10 Life expectancy at birth

- **Using the WorldMapper app:** The WorldMapper website (worldmapper.org) and app have been devised by geographers to produce cartograms that can show the relative size of any variable for all countries of the world. The user can select any one of more than a hundred variables, and immediately see a cartogram where the size of each nation is shown relative to all other nations for that variable. Such a resource is a very useful starting point or comparative source of data in an inquiry into human wellbeing.

- **Using the GapMinder app:** The GapMinder website (gapminder.org) and app allow the user to see a dot graph of the correlation between any two variables and, more importantly, see how both the variables change over time. By selecting different variables and tracing the pattern of different nations over time, global changes and local variations are vividly shown.

Content descriptor: Reasons for spatial variations between countries in selected indicators of human wellbeing (ACHGK077)

This unit builds on the skills of reading the statistics and maps to develop an inquiry into the reasons for such great variations. The influences of population growth, resources, government, history, gender and education should all be examined.

Core knowledge:

- Reasons for spatial contrasts in wellbeing

- Differences in wellbeing because of gender

- Population growth, resources and human wellbeing

Inquiry and skills:

- Interpreting and analysing multi-variable data such as population pyramids

- Making generalisations and inferences from data

Case studies:

- **Examining reasons for spatial variations in human wellbeing**: After looking at statistics and maps of the great variations between human wellbeing in different nations, suggestions about the reasons for these differences can be put forward and examined. Factors such as resources, population, government priorities and past history are among the factors that could be debated.

- **Inquiry into differences in wellbeing because of gender:** An inquiry into a variety of nations can extend the skills of data interpretation and representation. For any chosen nation, the following pieces of information should be extracted from databases on reputable websites: total population, growth rate, population pyramid showing percentages of males and females at each age cohort, death rate of babies, main causes of deaths for males and females, average incomes of males and females, and main occupations for males and females.

- **Causes and effects of fast population growth:** Students need to learn the meaning of the terms crude birth rate, crude death rate, population increase rate, infant mortality rate, fertility rate, population projections, population pyramids, population density, birth control and life expectancy. The use of these statistics forms the basis of an inquiry into the causes of fast population growth. The causes of the differences between the population growth in different countries can be examined in terms of resource base, government policies, education levels, dietary deficiencies and health services.

Content descriptor: Issues affecting development of places and their impact on human wellbeing, drawing on a study from a developing country or region in Africa, South America or the Pacific Islands (ACHGK078)

The concept of 'development' is quite complex. Statistics can show some of the facets of development, but there are many contrasting political views about it. A case study of a chosen country can highlight many of the issues connected to development.

Core knowledge:

- What is development, and what are its effects?
- Change through time in development

Inquiry and skills:

- Collecting and recording relevant geographical information
- Applying geographical concepts to synthesise information

Case studies:

- **The many faces of development:** Asking any group of people what they understand 'development' to mean will bring a wide variety of responses. Inquiring further into the meaning of economic development/social development/colonial development/sustainable development should extend ideas and produce more questions.

- **What are the most important development issues?** Statistics on income per capita, death rates, infant mortality, education achievement, doctors per capita, food availability and other measures of development can provide the raw material for learning about the ways in which development can be measured and compared. Positive and negative correlations can be highlighted, and decisions about which statistics best measure development can be debated. The statistics should be looked at in tables, bar graphs and scatter graphs.

- **Case study of a developing country:** Possible case studies could be Namibia or Ghana in Africa; any country in South America; or Fiji, Kiribati, Tonga or the Solomon Islands in the Pacific. The questions posed above could be applied to whichever country is chosen for a case study.

Content descriptor: Reasons for, and consequences of, spatial variations in human well-being on a regional scale within India or another country of the Asia region (ACHGK079)

India is a country of great diversity of development and human wellbeing. It is also a democracy that is becoming a major world power, and information about regional variations in its society is readily available.

Core knowledge:

- Contrasts in human wellbeing between different regions of a country

- Factors influencing human wellbeing

Inquiry and skills:

- Representing spatial distribution of geographical phenomena

Case studies:

- **Contrasts in human wellbeing between different regions of India:** India is a suitable case study for an understanding of contrasts in wellbeing because of its size, global importance and the availability of statistics and information. Specific states in India can be compared and contrasted for their level of wellbeing, based on statistics and researched information. Levels of life expectancy, literacy, income per capita and population increase can be compared between states such as Kerala, Maharashtra, Andhra Pradesh and Bihar.

- **Factors influencing human wellbeing in India:** Factors such as natural hazards (floods, tropical cyclones, droughts, landslides), population growth, corruption and political instability can be examined in general, and then related to different regions. The variation of these factors between regions is large.

Content descriptor: Reasons for, and consequences of, spatial variations in human wellbeing in Australia at the local scale (ACHGK080)

Although Australia has one of the highest indices of human wellbeing, it is obvious to students of this age group that there are great variations. Spatial variations include the differences between living in the city and the country. Other spatial variations can easily be observed between different suburbs in any of the large cities. The reasons for these differences are complex, and they have significantly different effects on different people.

Core knowledge:

- Spatial differences in the wellbeing of Aboriginal and Torres Strait Islander peoples

- City versus country variations

Inquiry and skills:

- Inquiring into relevant statistics to support suppositions

Case studies:

- **City versus country:** A list of what are perceived as good and bad points in a capital city and a country town can be a starting point for an analysis of people's wellbeing in these contrasting locations. The list might include accessibility of services, crowds, housing costs, space, recreation facilities, job availability, commuting to work, air pollution and many others. Then, details about the factors on the list can be checked for a city versus a country town.

- **Spatial differences in the wellbeing of Aboriginal and Torres Strait Islander peoples:** A starting point could be a map of distribution of the population of Aboriginal and Torres Strait Islander peoples across Australia. From this, there could be an inquiry into the wellbeing of Aboriginal and Torres Strait Islander peoples in large cities compared with rural areas. Other inquiries could focus on the Aboriginal and Torres Strait Islander peoples of contrasting rural areas such as Central Australia and the Kimberley.

Content descriptor: The role of international and national government and non-government organisations' initiatives in improving human wellbeing in Australia and other countries (ACHGK081)

Wealthy nations such as Australia see that they have an obligation to help poorer nations in many different ways. Global organisations such as the World Bank, non-government organisations (NGOs), individual nations and wealthy philanthropists all provide aid in various forms. Diverse opinions exist about the success of these initiatives, the source of funds, the types of projects and the commitments of wealthier nations.

Core knowledge:

- Effective ways of improving human wellbeing

- Should Australia spend more money on helping developing nations?

Inquiry and skills:

- Evaluating sources for their reliability, bias and usefulness

- Presenting findings, arguments and explanations

Case studies:

- **The work of World Vision, Oxfam, Médecins Sans Frontières (or any other global NGO):** Websites and material advertising for donations will provide the basic information about an NGO. However, further inquiry is needed to look beyond the advertisements to examine some of the projects it has have initiated, and to evaluate the success of its operations.

- **Should Australia spend more money on helping developing nations?** Australia spends about $4 billion a year on overseas aid, which is approximately 0.21 per cent of our gross domestic product (GDP) (World Vision, 2021). Surveys often indicate that people believe we spend a higher percentage, and that we should spend less. This provides the basis for an interesting debate or a student presentation based on facts and figures.

Short-answer questions

1. Which statistics and mapped data can be most effective in indicating differences in human wellbeing?

2. What is your understanding of the term 'development'? Contrast a narrow and a wide view of 'development'.

3. At what scales does the curriculum suggest that contrasts in human wellbeing be studied?

Unpacking the core knowledge and suitable case studies for Years 11 and 12

Geography courses in Years 11 and 12 vary from state to state. Although the Australian Curriculum: Geography gives a course outline for four units of work, each state has modified versions of these units, or locally developed units. In addition, each state has its own assessment authority, which manages the assessment standards and student achievement results. These authorities produce detailed guidelines for teachers and students about their courses, the assessment requirements and the resources for teaching and learning.

Nevertheless, there are key similarities and agreed approaches to senior geography across the states. The rationale and aims for senior geography are firmly based on those for Foundation to Year 10 geography. The same seven concepts used in earlier years are the basis of the courses, and each unit includes the same two strands of Knowledge and

Understanding, and Inquiry and Skills. Students have to master a wider range of geographical contexts and a more diverse and sophisticated range of geographical tools and skills. The General Capabilities and Cross-Curriculum Priorities are still an important emphasis within the courses.

Table 2.1 is a simplified summary of the topic names for the senior curriculum in each state in 2021. In most cases, there is a choice between or within the topics listed. Teachers and students should consult their state website curriculum for the details of requirements of their course.

Because the units of study are different in every state, as shown in Table 2.1, the units of the Australian Curriculum for Senior Geography have been chosen as an example to illustrate how the core knowledge, inquiry and skills can be developed with appropriate case studies.

Year 11

Unit 1: Natural and Ecological Hazards

This unit is related to similar topics in the senior geography curricula of Victoria, South Australia and Western Australia, and the points shown below can be used in interpreting the specific curricula of these states.

Core knowledge:

- The nature and types of atmospheric, hydrological, geomorphic and ecological hazards

- The factors contributing to risk and hazard – both physical and human

- The spatial distribution and scale of hazards of different kinds

Inquiry and skills:

- Using spatial technologies to illustrate the location and scale of hazards

- Interpreting tabular and mapped data

- Analysing patterns and trends

- Making predictions based on data

- Interpreting weather and climate maps for information about hazardous conditions

Case study:

- Depth study of one natural hazard and one ecological hazard:

    - The nature and causes of the hazard

    - Magnitude, frequency, spatial distribution and effects of the hazard

    - The activities of people, which contribute to the hazard

Table 2.1 Topics covered by individual states in senior geography

Australian Curriculum (ACARA 2020c)	Natural and ecological hazards	Sustainable places	Land cover transformations	Global transformations	
Queensland (QCAA 2021)	Managing the natural environment • Responding to natural hazards • Managing catchments	Social environments • Sustaining communities • Connecting people and places	Resources and the environment • Living with climate change • Sustaining biodiversity	People and development • Feeding the world's people • Exploring the geography of disease	
Victoria (VCAA 2021)	Hazards and disasters • Characteristics of hazards • Response to hazards and disasters	Tourism • Characteristics of tourism • Impact of tourism	Changing the land • Land use change • Land cover change	Human population – trends and issues • Population dynamics • Population issues and challenges	
New South Wales (NESA 2021)	Biophysical interactions	Ecosystems at risk	Urban places	Global challenges	People and economic activity
Western Australia (SCSA 2021)	Natural and ecological hazards	Planning sustainable places	Global environmental change	Global networks and interconnection	

Table 2.1 (*cont.*)

Australian Curriculum (ACARA 2020c)	Natural and ecological hazards	Sustainable places	Land cover transformations	Global transformations	Social and economic change
South Australia and Northern Territory (SACE 2021)	Hazards • Natural hazards • Biological and human-induced hazards	Sustainable place • Rural and/or remote places • Urban places • Megacities	Environmental change • Ecosystems and people • Climate change	Contemporary issues • Local issues • Global issues	Social and economic change • Population change • Globalisation • Transforming global inequality
Tasmania (TASC 2021)		Sustainable places • Urbanisation • Rural/urban interconnections • Population distribution in Australia • Population and economic change in Australia		Challenges facing places • Challenges for rural and remote places in Australia • Challenges for urban places in Australia	
ACT (BSSS 2021)	Natural and ecological hazards	Sustainable places	Land cover transformations	Global transformations	

- Physical and human factors that make some places more vulnerable

- Risk-management policies and procedures

Short-answer questions

1. List specific types of hazards (e.g. earthquake, tornado, drought, feral animals) that fit into the general categories of hazard listed above.

2. Suggest locations that have had major hazard events, which will provide data about causes, effects and responses.

3. Suggest ways in which the use of spatial technologies could be the basis of analysis of patterns, trends, magnitude and frequency of a particular hazard.

4. Suggest some learning experiences and particular examples of hazard events that could show how the actions of people can increase the risk of a natural hazard (e.g. building houses in bushfire areas, cutting down trees on steep slopes prone to landslide).

5. Suggest examples of building laws, precautions and defences that have lowered particular hazard risks.

Unit 2: Sustainable Places

This unit is related to similar topics in the senior geography curricula of Tasmania, South Australia and New South Wales, and the points shown below can be used in interpreting the specific curriculum of these states.

Core knowledge:

- The process of urbanisation and its contribution to world population growth

- The influence of urbanisation on both urban and rural places

- The concepts of push and pull in migration

- The economic and environmental links between urban and rural places

- The spatial distribution of metropolitan, regional and rural places in Australia and their changing populations and functions

- Challenges facing rural and metropolitan places in Australia

- Challenges facing megacities in developing countries

Inquiry and skills:

- Collecting and interpretating of statistical data relating to populations, wealth, jobs, migration and housing in rural, regional and metropolitan places

- Analysing and interpreting population pyramids for different locations and regions

- Analysing patterns leading to reasoned generalisations

- Using fieldwork and secondary sources to investigate changes and challenges to a local place

- Using atlas maps (both paper and digital) to locate places

Case studies:

- Depth study of challenges facing a place in Australia:

  - Inquiry into the challenges facing a particular place (e.g. migration of a particular age group, lack of facilities for certain groups of people)

  - Identification of strategies being used to address the challenges (e.g. laws, incentives)

  - Evaluation of the effects of strategies on sustainability and liveability of the place

- Depth study of challenges facing a megacity in a developing country:

  - Inquiry into the challenges facing a megacity (e.g. Jakarta, Sao Paulo, Bangkok, Kolkata)

  - Identification of strategies being used to address the challenges (e.g. housing, jobs, transport systems, infra structure)

  - Evaluation of the effects of these strategies on the sustainability and liveability of the megacity

Short-answer questions

1. Which infographics would you decide to use in order to best illustrate the growth of urbanisation and megacities in the world?

2. How would you teach about rural places differently if you were teaching in an urban school compared with teaching in a rural school?

3. Suggest some examples of economic and environmental links between rural and urban places in Australia.

4. What are some of the spatial patterns and issues of Australian capital cities that are smaller scale versions of those in overseas megacities?

Year 12

Unit 3: Land Cover Transformations

This unit is related to similar topics in the senior geography curricula of Victoria and New South Wales, and the points shown below can be used in interpreting the specific curriculum of these states.

Core knowledge:

- Types of land cover and biomes: forests, cropland, rangelands, pasture, urban land

- Effects of population growth, affluence, advances in technology on land cover and biodiversity

- Traditional land management practices of Aboriginal and Torres Strait Islander peoples and their influence on land cover

- The effects of climate change on land cover

- The comparative effects of indigenous and introduced species of plants and animals on biomes and land cover

- The concept of anthropogenic biomes

Inquiry and skills:

- Recognising types of land cover: forests, cropland, rangelands, pasture, urban land from aerial photographs

- Using fieldwork to map, photograph and record examples of differences and changes to land cover

- Interpreting topographic maps

- Analysing digital aerial photographs taken on different dates to show land cover change over time

Case study:

- Depth study of the interrelationship between land cover change and changes in either global climate or biodiversity:

  - Causes and projected impacts of climate change

  - Effects of climate change on vegetation, ice sheets, coastal features

  - Causes and impacts of declining biodiversity

  - Effects of biodiversity loss on ecosystems, species and genetic diversity

  - Inquiry into a local initiative relating to climate change or biodiversity loss

Short-answer questions

1. What would be a viable fieldwork activity concerning land cover in a rural area with agriculture, small towns and natural vegetation?

2. In what ways could you devise a fieldwork activity for an urban area that would develop some of the skills listed above, focusing on small scale differences in land cover (such as lawns, gardens, buildings, roads, trees, car parks)?

3. Which areas do you know of that would be suitable locations for a virtual field study using digital maps?

Unit 4: Global Transformations

This unit is related to similar topics in the Senior Geography curricula of Victoria and New South Wales, South Australia and Western Australia, and the points shown below can be used in interpreting the specific curriculum of these states.

Core knowledge:

-   Changes in the spatial distribution of production and consumption of goods and services

-   The impact of new technologies in telecommunications and transport

-   The power of world cities: financial, cultural, intellectual

-   The growth of China and India as world powers

-   The cultural influence of Europe and the United States across the world

-   The disintegration of aspects of globalisation caused by pandemics and changing political power

Inquiry and skills:

-   Analysis of population growth statistics for newly developing regions

-   Identifying trends and patterns in global integration via technology (e.g. SMS messaging, social media, streaming services, internet retail purchasing)

-   Communicating ideas, issues and arguments about the possible future of globalisation, including both positive and negative effects

Case study:

-   International economic integration of an energy resource, food commodity, manufactured commodity or service:

    -   Changes in spatial patterns and the reasons for them

- The role played by changes to transport and technology, and changes to import and export barriers

- The effects of these changes on people, places and environment

- Likely future changes – short and long term

- The consequences of different responses of people to the effects of globalisation

Short-answer questions

1. If you were choosing a particular energy resource, food, manufactured commodity or service for your class for the in-depth case study, what would you choose and why? Give reasons that relate to resources, school location, student group and interests.

2. Students who have always lived in a world that has been globally connected by the internet may have different expectations of the future than those of an older generation. What differences would you expect?

3. With which aspects of globalisation do you think students are familiar, and with which aspects would they be unfamiliar? How would you take this into account in preparing lessons for this unit?

Field study reports in senior geography

Every state has a requirement of some type that students must undertake fieldwork and present a field study that makes use of the fieldwork. This means that students need to become aware of the range of field techniques that they can use, be able to apply the chosen techniques accurately and use their findings to present a well-reasoned field study report.

Chapters 5 and 9 give teachers guidance on fieldwork. *Geography Fieldwork Unlocked* (Kleeman 2019) provides examples of fieldwork for each of the curriculum units from Years 7 to 10.

Conclusion

This chapter specifically covered content and case studies in the teaching of the Australian Curriculum: Geography. Every teacher must ensure they understand the key ideas (content) of each of the content descriptors and how the use of case studies (in the widest sense) helps in the teaching of this.

This chapter also showed that a variety of methodologies is essential in the classroom, fieldwork and homework. Therefore, as well as interpreting the content in the curriculum, various different techniques that can be used to teach particular content and case studies have been included. The wide range of teaching techniques provides diversity that encourages teachers to make choices about which methods to use, given all the possible variations in classroom dynamics, location and resources.

Bringing it together

1. Which topics in secondary geography reflect the concept of the spiral curriculum as outlined by Jerome Bruner (1960)?

2. Choose some examples from the Australian Curriculum: Geography that reflect the constructivist view of knowledge, and outline how they reflect this view.

3. After reading the case studies in this chapter, make a list of the different teaching techniques suggested in them.

4. From the many descriptions of case studies in the chapter, choose one that uses teaching techniques you would like to apply to a different area of content. Briefly outline what you would like to do.

5. Which units (in any year level from 7 to 12) contained geographical content material with which you were unfamiliar? Suggest further questions that you would like answered about that area of geography.

6. Choose any of the units described and plan a sequence of inquiry, incorporating skills and content described, but using different case studies to develop learning.

References

ACT Board of Senior Secondary Studies [BSSS] (2021) *Frameworks: Curriculum documents for Years 11 and 12*, ACT Government, retrieved from www.bsss.act.edu.au/curriculum/Frameworks

Australian Curriculum, Assessment and Reporting Authority [ACARA] (2020a) *F–10 Curriculum: Humanities and Social Sciences*, Canberra: ACARA, retrieved from www.australiancurriculum.edu.au/f-10-curriculum/humanities-and-social-sciences/hass

——(2020b) *F–10 Curriculum: Geography*, Canberra: ACARA, retrieved from www.australiancurriculum.edu.au/f-10-curriculum/humanities-and-social-sciences/geography

——(2020c) *Senior Secondary Curriculum: Geography*, Canberra: ACARA, retrieved from www.australiancurriculum.edu.au/senior-secondary-curriculum/humanities-and-social-sciences/geography

Bruner, J. (1960) *The process of education*, Cambridge, MA: The President and Fellows of Harvard College.

Kleeman, G (2019) *Geography fieldwork unlocked*, Sydney: Australian Geography Teachers Association.

NSW Education Standards Authority [NESA] (2021) *Geography Stage 6 Syllabus*, Sydney: NESA, retrieved from www.educationstandards.nsw.edu.au/wps/portal/nesa/11-12/stage-6-learning-areas/hsie/geography

Office of Tasmanian Assessment, Standards & Certification [TASC] (2021) *Geography*, Hobart: TASC, retrieved from www.tasc.tas.gov.au/students/courses/humanities-and-social-sciences/ggy315120

Queensland Curriculum & Assessment Authority [QCAA] (2021) *Geography*, Brisbane: Queensland Government, retrieved from www.qcaa.qld.edu.au/senior/senior-subjects/humanities-social-sciences/geography

School Curriculum and Standards Authority [SCSA] (2021) *Geography: Syllabus and support*, Perth: Government of Western Australia, retrieved from https://senior-secondary.scsa.wa.edu.au/syllabus-and-support-materials/humanities-and-social-sciences/geography

South Australian Certificate of Education [SACE] (2021) *Geography: Stage 1 and 2*, Adelaide: SACE, retrieved from www.sace.sa.edu.au/web/geography

United Nations, Department of Economic and Social Affairs, Population Division (2019) *World population prospects 2019: Highlights*, Geneva: United Nations.

Victorian Curriculum and Assessment Authority [VCAA] (2021) *Study Design: Geography*, Melbourne: VCAA, retrieved from www.vcaa.vic.edu.au/curriculum/vce/vce-study-designs/geography/Pages/Index.aspx

World Vision (2021) *Why does Australia give aid?*, Sydney: World Vision Australia, retrieved from www.worldvision.com.au/get-involved/advocacy/australian-aid

PART 2
GEOGRAPHICAL SKILLS

CHAPTER 3

The graphicacy of geography

Rebecca Nicholas

Learning objectives

By the end of this chapter, you should be able to:

- understand what graphicacy is, why it is an essential literacy and what it looks like in the Australian Curriculum: Geography

- explore the different types of graphicacy

- identify strategies to support the explicit teaching of graphicacy in the geography classroom

Introduction

In our world of the 24-hour news cycle, big data and social media, the volume of information being generated at a global scale is unprecedented. We are required to consume and make sense of this information on a daily basis. Information is mostly presented in graphic form, and in many instances includes a spatial element. Students in the twenty-first century must be critical and creative encoders and decoders of all this information. The relevance of graphicacy – the ability to understand and present information in graphic form – and the literacies associated with this should be an essential capability in curriculum design. Unfortunately, graphicacy is considered the 'forgotten' literacy, particularly in secondary and tertiary education. Geography provides a structured approach for educators to develop skills and embed 'graphicacy', along with literacy and numeracy from Foundation to Year 12.

This chapter will explore the essential role played by geography in developing a student's 'graphicacy'. It will look at the relationship between graphicacy, visual literacy and visual thinking in geography, and explore why this is an essential part of the geography curriculum. The differing types of graphicacy will be investigated, along with strategies to support the effective teaching of graphicacy in the classroom.

What is graphicacy?

Graphics predate numbers and writing as the earliest form of communication. Graphicacy is concerned with the human ability to make and interpret meaningful marks (Danos 2014, p. 13). Graphics have the ability to catch our attention and are an essential tool for communication (Poracsky, Young & Patton 1999).

Over the last 100 years, there has been an evolution in the way we communicate from 'typographic' to 'graphic'. Roux (2009) found that the vocabulary of the average 14-year-old had dropped from 23 000 words in 1950 to about 10 000 words in 1999, and has gradually been decreasing since then. This is not due to students being less articulate: communication has become multimodal and we live in an environment surrounded by visual media. Social, economic and technological developments have changed the traditional concept of what it means to be literate (De Jager 2014, p. 1). People now communicate more frequently with either images and text or images alone. Emojis and social media platforms are an indication of this. The well-known phrase 'A picture is worth one thousand words' is no more relevant than now.

Graphicacy is the ability to understand, read and create visual images as a means of communication (Danos 2014, p. 34). When considering 'graphics', we immediately think of photographs or pictures. However, graphicacy is more complex than this. Graphics includes all types of maps, graphs and diagrams. These graphic images are considered more complex as they can communicate quantitative values and trends, illustrate spatial patterns and allow interconnections to be recognised and analysed. Graphicacy moves 'beyond just being visually literate and instead combines mathematics, statistical analysis, geographic interpretation, [spatial thinking] and graphic design' (Hall & Russac 2011).

Students need to become 'smart' encoders and decoders of graphics. Any form of literacy – written, numerical or graphical – requires higher level thinking skills such as extrapolation, analysis and synthesis to both comprehend and create. While graphicacy does not replace other types of literacy, it does provide a critical way of thinking that brings text and numbers together. This is why Balchin and Coleman (1965, p. 85) argue that 'literacy, numeracy, oracy and graphicacy are the "four aces" in the pack of education. If any of these is left out of the pack, education is incomplete.'

Pause and think

Consider the news reporting of local, state or federal elections. Think of two ways in which strong graphicacy (graphical literacy) supported you in interpreting data associated with the political event occurring at the time.

Graphicacy = visual literacy + visual thinking

Graphicacy includes both visual literacy and visual thinking. Visual literacy involves the ability to comprehend, make meaning and communicate through visual means. Students who are visually literate understand how images and language work together to present ideas and information in texts (Kleeman 2017). This is the 'decoding' part of literacy, and it involves students comprehending the visual text and then deconstructing for meaning. By explicitly teaching visual literacy skills in the geography classroom, students are better able to analyse data, as well as describe geographic relationships, patterns and trends.

Visual thinking involves the encoding of data or information to create a visual representation. This ability to 'encode' information has been described by Fry (1981, p. 388) as the other half of graphical literacy. Creating a graphic, be it a map, graph, field sketch or annotated diagram involves knowledge and higher order thinking skills. For a student to create a graphic they need to understand the data, develop a clear purpose, follow graphic or cartographic conventions and ensure their graphic is not manipulating the data when communicating. In the Australian Curriculum: Geography this is referred to as 'representing' data.

Figure 3.1 describes the relationship between graphicacy, visual literacy and visual thinking. The diagram highlights the higher level cognitive processes involved in both 'encoding' and 'decoding' graphics. In this example, Marzano and Kendall's (2007)

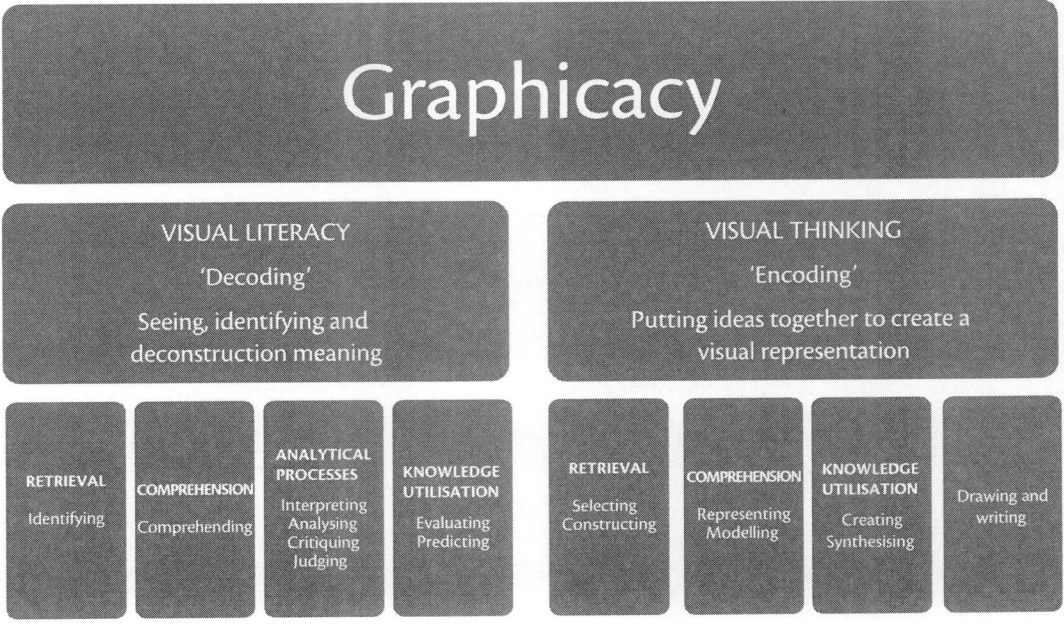

Figure 3.1 The relationship between graphicacy, visual literacy and visual thinking and the cognitive processes associated with these skills (Marzano and Kendall's (2007) four levels of cognitive process are noted in black)

Source: Adapted from Hall & Russac (2011); Marzano & Kendall (2007); Wilmot (1999).

taxonomy of educational objectives has been used to categorise cognitive processes. Teachers should not assume that students have the necessary skills (reading, writing, numeracy and drawing) or have developed the cognitive processes, particularly analytical processes and knowledge utilisation, to either decode or encode images when mapping graphicacy across the curriculum. These skills require explicit teaching, time and space in the curriculum, and this should be considered when planning units and assessment.

Pause and think

Consider a lesson that involves teaching students to draw and then interpret a simple bar graph. Using the headings of visual literacy and visual thinking, identify the specific cognitive processes and skills required for a student to complete this task. Reflect on the usual time allocated in the lesson to develop this form of graphicacy. Have students been given time and space to complete this activity, considering the cognitive demand of this task?

Why is graphicacy an important part of the geography classroom?

Geography is a uniquely visual discipline. Fundamentally, the communication of spatial information cannot be conveyed adequately by verbal or numerical means alone (Balchin 1972). For example, when asked to provide directions to a specific location, a sketch map is drawn to support the verbal instruction. Due to the nature of the discipline, graphicacy is an essential literacy that is developed in the geography classroom.

Geography students are exposed to a variety of images, graphs, maps, diagrams and models, which means they become critical consumers of visual media on the completion of their education. Additionally, future careers will rely on students being able to encode and decode data and information with ease. Geography is a subject that develops these skills in students. The increased use of ICTs in the geography classroom means that visual representations can be created, reproduced and edited with comparative ease. The use of spatial technologies, online databases and data-manipulation software such as Microsoft Excel means that by the time students finish their studies in geography, they will be able to successfully 'represent' data to communicate information.

What does graphicacy look like in the Australian Curriculum: Geography?

Graphicacy, in the form of visual literacy, is found in the Literacy general capability of the ACARA curriculum documents. Specifically, the Literacy learning continuum

(Figure 3.2) includes visual literacy as one of the six interrelated elements. Visual knowledge is also included in the general capability of Literacy.

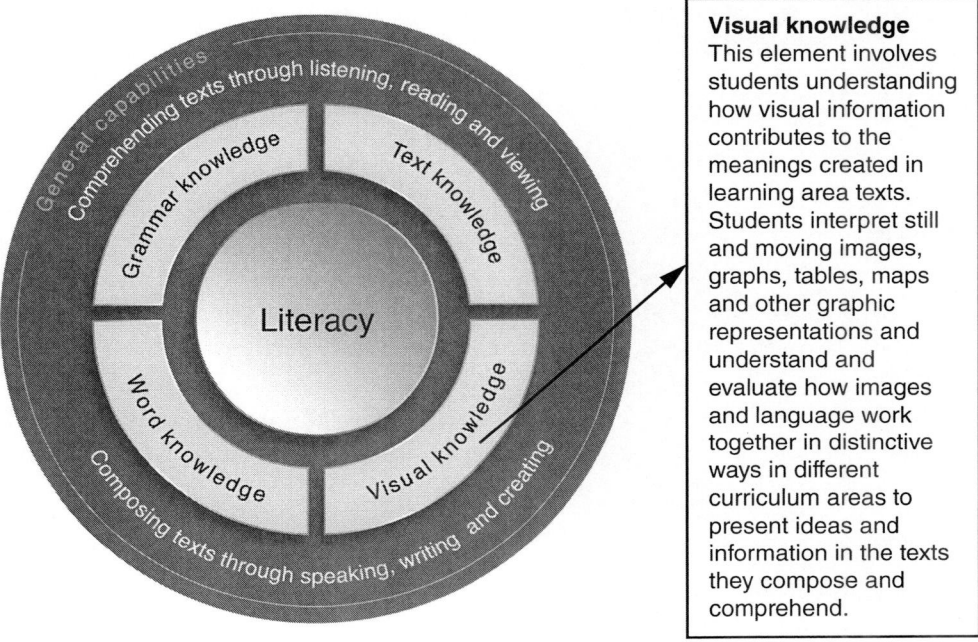

Visual knowledge
This element involves students understanding how visual information contributes to the meanings created in learning area texts. Students interpret still and moving images, graphs, tables, maps and other graphic representations and understand and evaluate how images and language work together in distinctive ways in different curriculum areas to present ideas and information in the texts they compose and comprehend.

Figure 3.2 The Literacy continuum includes two overarching processes (comprehending and composing texts), as well as areas of knowledge that apply to both processes. Visual knowledge is included here.

Source: Adapted from ACARA (2020a).

An investigation of the Australian Curriculum: Geography indicates that graphicacy is clearly developed from Foundation to Year 12. Students are expected to use a wide range of texts, both written and graphic. Graphicacy is more obvious in the Inquiry and Skills strand. In this strand, students are required to:

- collect, record, evaluate and represent geographic data

- interpret, analyse and make conclusions using geographic data; this includes 'identifying order, diversity, patterns, distributions, trends, anomalies, generalisations and cause-and-effect relationships' (ACARA 2020b)

- communicate geographic data collected in the field, or in a research task, using both text and graphic forms.

While the curriculum provides a framework for the development of graphicacy, this may not be the case at a school level. An audit of assessment and unit plans, particularly in a high school context, would show any gaps or areas of development in terms of graphicacy.

The Connection box on mapping graphicacy across the Australian Curriculum: Geography provides information on how such an audit could be conducted at a school level.

CONNECTION

Mapping graphicacy across the Australian Curriculum: Geography

Table 3.1 provides an overview of graphicacy in the Australian Curriculum: Geography. It should be noted that both the content descriptors and elaborations were used to collect this data. Additionally, the type of graphicacy was only highlighted when the graphic (such as map, graph) was specifically referred to. Teachers would include specific types of graphicacy across stages of learning to suit context and topic, even if it is not explicitly mentioned in the Inquiry and Skills strand. Often the term 'a range of formats' has been used, particularly in the secondary curriculum, to allow for flexibility for the school curriculum context. The Humanities and Social Sciences (HASS) curriculum was used when mapping in the primary years.

Questions

1. Predict what graphicacy would look like for Year 11 and 12 students in your state or territory, considering the progression indicated in Table 3.1. What would you expect to see in the curriculum documents? Check your thinking by accessing the senior geography curriculum documents for your state or territory.

2. Access a geography text commonly used in schools in your state for one unit of work in one of the phases of learning (e.g. Australian Curriculum Year 7 – Unit 1: Water in the World). Conduct an audit of this text to determine if it meets the requirements of graphicacy in terms of decoding and encoding, as well as the differing types of graphicacy.

Short-answer questions

1. In your own words, define the term 'graphicacy' and explain what it looks like in the geography classroom.

2. What are the two types of thinking that contribute to graphicacy?

3. Why would graphicacy be associated with higher order cognitive processes?

4. How has technological change resulted in an increased importance of graphicacy in the geography classroom?

Table 3.1 A continuum of graphicacy in the Australian Curriculum: Geography, with a focus on visual literacy and visual thinking as well as the types of graphicacy

Type of graphicacy	Foundation – Year 2	Years 3 & 4	Years 5 & 6	Years 7 & 8	Years 9 & 10
Visual literacy 'decoding'	• Interpret data on pictures (*) • Interpret data on maps (^)	• Interpret data in different formats to identify and describe distributions and simple patterns	• Interpret data and information displayed in a range of forms to identify, describe and compare distribution, patterns and trends and to infer relationships	• Interpret aerial images (*) • Interpret weather, isopleth, topographic, political, thematic maps (^) • Interpret digital terrain models, cross-sections or block models (~~) • Interpret geographical data to identify and propose explanations for spatial distribution patterns, trends and infer relationships • Use digital maps and overlays to observe spatial associations (^) • Draw conclusions based on analysis of data.	• Interpret satellite images (*) • Identify relevant layers using GIS to analyse spatial data (^) • Critically analyse images for their meaning and significance • Interpret and analyse multi-variable data and other geographical information to make generalisations and inferences, propose explanations for patterns, trends, relationships and anomalies, and predict outcomes

Table 3.1 (*cont.*)

Type of graphicacy	Foundation – Year 2	Years 3 & 4	Years 5 & 6	Years 7 & 8	Years 9 & 10
Visual thinking 'encoding'	• Record information and data on labelled maps (^) • Sort and record information and data in tables (^^) • Represent and locate features on models (~~)	• Sort and represent data in simple: – Graph format (**) – Map format (^) – Table format (^^)	• Organise and represent data using appropriate conventions: – Graphs (**) – Tables (^^) • Organise and represent data in large- and small-scale maps using appropriate conventions (^)	• Represent data in a range of appropriate forms, e.g. climate graphs, compound column graphs, population pyramids (**), tables (^^) • Construct maps at different scales using cartographic conventions (^) • Create a map showing spatial distribution (^) • Construct field sketches • Create annotated diagrams of waterflows and landforms (~)	• Represent multi-variable data in a range of appropriate forms • Construct graphs to show relationships between data, e.g. scatterplots (**) • Represent spatial distribution of geographical phenomena by constructing special purpose maps that conform to cartographic conventions, e.g. choropleth maps (^) • Create a map using spatial technologies (^) • Construct field sketches and annotated diagrams (~)

Key: * Images ** Graphs ^ Maps ^^ Table ~ Diagrams ~~ Models

Source: Adapted from ACARA (2020c, 2020d).

Types of graphicacy

Researchers have categorised types of graphics in a variety of ways. Danos (2014, p. 37) concludes that images included in graphicacy could be categorised as:

- information conveyed in a directly representative form (sketches, photographs, drawings)

- information conveyed in an abstract form (diagrams, plans, graphs, tables, maps)

- information conveyed in a variety of forms (non-textual, two-dimensional, three-dimensional).

Fry (1981) presents a similar breakdown of categories, but is more specific in the explanation of these to suit education. This section of the chapter will provide an overview of the differing types of graphicacy through the lens of the Australian Curriculum: Geography.

CONNECTION

Fry's taxonomy of graphs

In 1981, Edward Fry published an article on graphical literacy in which he developed a taxonomy of graphs as it related to curriculum. This taxonomy was developed to broaden the term 'graph' beyond the scope of a bar, pie or line graph. Fry created six categories of graphic displays. This taxonomy provides a comprehensive overview of the differing types of graphs or graphics to which students should be exposed as part of their education:

1. Lineal – sequential data

   - Simple

   - Multiple

   - Complex

   - Hierarchy

   - Flow

   - Process

   - Sociogram

2. Quantitative – numerical data

   – Frequency polygon

   – Bar graph

   – Pie graph

   – Complex

3. Spatial – area and location

   – Two dimensions

   – Three dimensions

4. Pictorial – visual concepts

   – Realistic

   – Semi-pictorial

   – Abstract

5. Hypothetical

   – Conceptual

   – Verbal

6. Omitted – details absent

   – High verbal

   – High numeric

   – Symbols

   – Decorative design

Questions

1. Which of the types of graphs in Fry's taxonomy would feature most often in a geography classroom? Why would this be the case?

2. Why would it be important to expose students to each category of the taxonomy over the course of their education?

3. Look across one or two year levels in the Australian Curriculum: Geography
    (e.g. Years 7 and 8). Identify the different types of graphs that are suggested in the
    curriculum (including the standard elaborations). Are there any gaps?

Images

The use of images is an essential element of pedagogy in geography. The rationale of the Australian Curriculum: Geography states that the subject should 'inspire curiosity and wonder about the diversity of the world's places, peoples, cultures and environments' (ACARA 2020e). In all effective geography classrooms, images are used to engage and motivate students.

Images illustrating geographical knowledge (e.g. landscapes) or concepts (e.g. change over time) support student learning. Images – particularly ground photos and field sketches – are also an essential skill in the collection of field data.

Images in geography include:

- ground photos
- aerial photos (vertical and oblique)
- satellite images including remotely sensed images.

Cognitively, students are often asked to identify and describe natural or human features, annotate images to explain geographical processes or compare satellite imagery to maps to analyse spatial or temporal change.

Improved access to technology in the classroom has meant that teachers and students can now access a plethora of images to support learning. The creation of spatial tools such as Google Earth and Google Street View mean that students have access to satellite imagery and 3D imagery from the global to the local scale, including at street level.

Pause and think

How have you used images when planning a lesson as part of your course? Think about whether you have considered investigating an image to develop knowledge and understanding, rather than providing text or notes to read.

These technological advancements have also meant that images are no longer used as descriptive illustrations. Students should be critical users of these images, understanding the importance of how they can be seen and used by particular viewers in particular contexts. The Connection box on analysing visual images provides a strategy for students to analyse images in this way.

Images are also used as a way of identifying stereotypes and misconceptions, particularly in human geography. A combination of statistical data and image visualisations in online tools such as 'Dollar Street' (gapminder.org/dollar-street) has meant that students can not only describe an image but also analyse it with quantifiable data.

CONNECTION

Analysing visual images – see, think, wonder

The following strategy may be used with students when critically viewing visual media in the classroom.

1. Select an image

Choose a photograph, cartoon, poster or other image (such as a satellite image). The image should reveal information about a particular time or reflect a particular perspective for the activity to provide enough discussion with students. Display the image on the screen and provide students with a copy of the image.

2. Ask the following questions

1. What do you see? What details stand out?

Ask students to label or annotate the image or list these. These must be 'on the page' observations that are 'point-out-able'.

2. What do you think is occurring? What makes you say that?

Students should make inferences and 'read between the lines' on the image. Students should write down their responses, as well as evidence from the image to support this thinking.

3. What does this image make you wonder? What broader questions does this image raise with you?

Between each question, give students time to think (up to five minutes), talk to their partner and then ask some pairs to share their thinking (think, pair, share strategy).

Following a discussion with the class, students could incorporate their knowledge and understanding of the topic studied to answer the third question in a 100–200-word paragraph.

Question

What types of images and graphics in a geography lesson could this strategy work with and why?

Graphs

Graphs are very common in the geography classroom as they effectively communicate geographic data. Graphs can pack high-density information into a small area and are the most effective way to recognise and interpret trends, patterns and relationships. Students should be exposed to a variety of simple and complex graphs across the course of their study, particularly when interpreting and analysing geographic data.

Types of graphs in geography

A large number of graphs should feature in the geography curriculum, including:

- bar and column graphs
- complex bar charts
- divergent bar charts
- histograms
- line graphs
- cross-sections
- pie charts
- climate graphs
- population profiles
- scatter graphs
- ternary graphs
- radar graphs
- development diamonds
- kite diagrams.

By Year 10, students should be familiar with both encoding and decoding a majority of these types of graphs. The additional web materials that support this chapter provide an overview of the types of graphs and links to the curriculum. This resource is useful when developing units and lesson plans that should incorporate graphing from Years 7 to 10. An overview of one of these graph types – population profiles – is provided in the Connection box.

CONNECTION

Learning experiences using population profiles in the geography classroom

What is a population profile?

Population profiles, also called population pyramids, are 'graphs that represent the age and sex composition of a population using bar graphs' (ACARA 2020f). A population profile is used to determine the age and sex structure of a population.

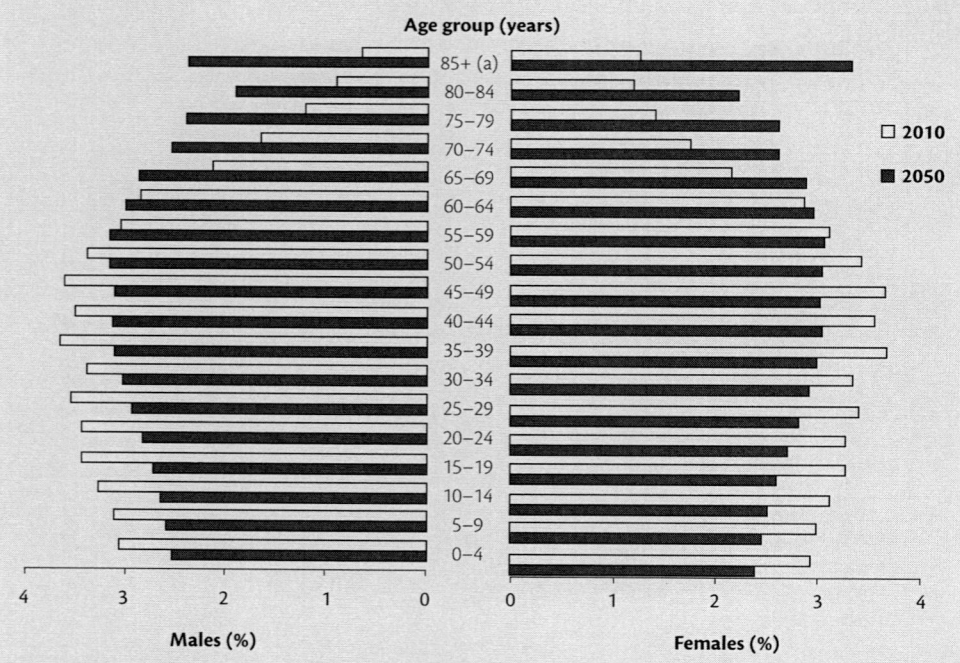

(a) The 85+ age group includes all ages 85 years and over and is not directly comparable to the other five-year age groups.
Source: Population Projections, Australia (3222.0).

Figure 3.3 A population profile for Australia in 2010 and a projection for 2050

Source: ABS (2018).

Where do population pyramids fit into the Australian Curriculum: Geography?

Population profiles could be included as a key graphicacy skill in the following units:

- Year 7 – Place and Liveability to represent secondary data collection in local fieldwork

- Year 8 – Changing Nations to support interpretation of data when analysing causes of urbanisation and the management of Australia's urban future

- Year 9 – Biomes and Food Security when interpreting and analysing global population projections

- Year 10 – Geographies of Human Wellbeing as part of the study of demographic data.

Sample learning experiences

Encoding population profiles

Year 7 students are provided with a table of age-sex data for their city as a table. Teachers model, step by step, the process and thinking involved in the creation of a population profile. Teachers should include graphic conventions (SALTS). The profile could be constructed using Microsoft Excel. A series of guiding questions are then provided by the teacher to interpret the profile for the suburb, and how this could impact on liveability.

Decoding population profiles

Year 10 students use various resources to interpret population profiles of various countries and categorise them (expansive, constrictive or stationary). Using their existing knowledge of the demographic transition model, they justify the location of each of the profiles on the stages of this model. Students then discuss the implications on future wellbeing for each of these selected countries.

Maps

Maps are a way of communicating information about places. Geographers use maps as a tool to locate places, understand patterns of natural and human features, and analyse relationships between these features.

Traditionally, maps in the geography classroom were centred around place-name geography. Considerably more time was spent on spatial literacy – identifying and remembering where places were on a map. Today, maps are essential tools used to interpret and analyse spatial distribution, patterns and relationships when using geographic data and information. Students should be exposed to a range of differing maps from simple to complex from Years 7 to 12. Students should also have the opportunity to use spatial and digital technologies to create and construct a variety of maps.

Types of maps in geography

Students will use the following types of maps when studying geography:

- physical map
- political map
- topographic map
- thematic map
- synoptic map
- precis map
- dot map
- choropleth map
- isoline map
- proportional symbols map
- cartogram
- flowline map.

By Year 10, students should be familiar with both encoding and decoding a majority of these types of maps. The additional web materials that support this chapter provide an overview of the types of maps and the link to the curriculum. This resource is useful when developing units and lesson plans that should incorporate graphing from Years 7 to 10. An overview of one of these map types – the choropleth map – is provided in the Connection box.

CONNECTION

Learning experience using choropleth maps in the geography classroom

What is a choropleth map?

Choropleth maps (e.g. Figure 3.4) show the density of objects or values in a given area by means of shading or patterns. The shading or pattern of an area changes according to the differing values. Choropleth maps use monochromatic shading, with the lower values in lighter shades and the highest values in the darkest shades.

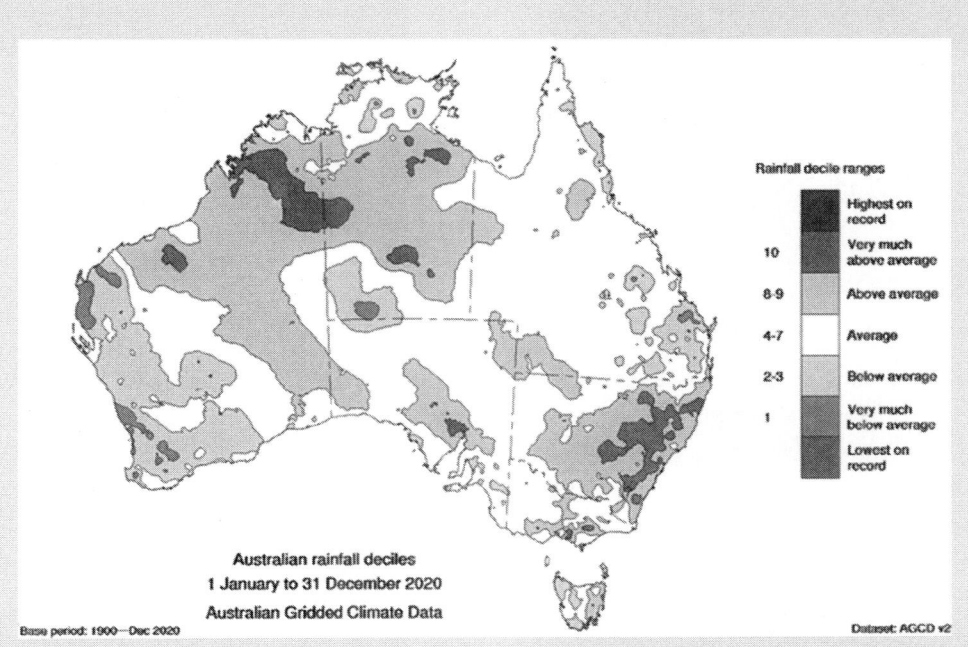

Figure 3.4 Choropleth map showing 12-monthly rainfall deciles for Australia between 1 January and 31 December 2020

Source: BoM (2021).

Where do choropleth maps fit into the Australian Curriculum: Geography?

Most units from Years 7 to 12 should include the use and interpretation of choropleth maps. For example, Year 7 – Water and the World could include a choropleth map activity that looks at the nature of water scarcity, and Year 8 – Changing Nations includes the use of choropleth maps to gather data on urbanisation and migration. By Year 9, students should be using tools such as Microsoft Excel, MapChart or spatial tools to create their own choropleth maps based on data provided in a learning experience.

Sample learning experiences

Teachers should use the PQE strategy, outlined in the web materials for this chapter, to support students when interpreting the patterns and relationships between data represented in choropleth maps. Most online data websites, including the World Health Organization, World Bank and Our World in Data, provide an option to show any selected data as a choropleth map.

Pause and think

Maps and graphs can be either simple or complex. Consider the types of graphs and maps that students are most frequently required to interpret or create. Which types dominate student learning – simple or complex? What does this mean for the graphical literacy of a student completing Year 12?

Diagrams

Diagrams are tools that show the arrangement of and relationship that exists between various parts of a process. Diagrams can range from simple to complex, and are used to simplify a concept, summarise information or support written information when communicating geographically. They can be created from data collected in the field, used to describe physical processes and to show complex relationships between concepts. The additional web material provides further detail on the differing types of diagrams and how these are used to support students in the geography classroom.

Short-answer questions

1. What are the five types of graphicacy as explained by Fry in his taxonomy?

2. How has the use of images changed in the geography classroom?

3. Graphs, maps and diagrams can be categorised as simple or complex. What is the difference between simple and complex in terms of thinking for a student?

Strategies to include graphicacy in the geography classroom

Teachers should not assume that students can read, interpret or know how to create graphics. Each time a graphic is used, there should be time allowed for the explicit teaching of strategies that develop a student's ability to decode and encode geographic information.

Encoding – strategies for representing graphical data

Generally, students find it difficult and challenging to create graphic and cartographic representations, particularly when they are required to select the best graphic to communicate the meaning of the data. To successfully create a graph, map or diagram, students are required to:

- find a suitable dataset
- determine the trend, relationship, pattern or story the data or information is telling
- synthesise this data
- determine the most appropriate graph, map or diagram to represent this information or data
- create the graph or map correctly following graphic or cartographic conventions (SALTS or BOLTSS)
- combine the graphic effectively with written text and other images when communicating.

Tufte's principles of graphical excellence provide a more sophisticated checklist to consider when creating and communicating with data.

CONNECTION

Tufte's principles of graphical excellence

Edward Tufte is Professor Emeritus at Yale University, where he taught courses in statistical evidence, analytical design and political economy. In his text, *The Visual Display of Quantitative Data* (Tufte 2001), he provides a simple overview of graphical excellence.

Graphical displays should:

- show the data
- avoid distorting what the data have to say
- present many numbers in a small space
- make large datasets coherent
- encourage the eye to compare different pieces of data
- reveal the data at several levels of detail, from a broad overview to the fine structure
- serve a reasonably clear purpose: description, exploration, tabulation or decoration
- be closely integrated with the statistical and verbal descriptions of the dataset.

Graphical excellence:

- is the well-designed presentation of interesting data

- consists of complex ideas communicated with clarity, precision and efficiency

- gives the viewer the greatest number of ideas in the shortest time with the least ink in the smallest space.

Questions

1. Tufte's principles of graphical excellence are quite sophisticated. Develop a checklist that could be used by high school students to support them in the creation of excellent graphics.

2. Find a graphic that provides data on the coronavirus deaths in 2020 in various countries around the world. Evaluate the effectiveness of this graph using Tufte's principles of graphical excellence. Does the graph meet these principles?

Understanding when one mode of graphic is more appropriate than another, and how to use each mode effectively, is essential knowledge and an important skill. Teachers should therefore focus on the use of modelling, think-alouds and examples to support students in the successful construction of graphics.

According to the Australian Curriculum: Geography, by Year 10 students should be able to represent data to communicate meaning by selecting the most appropriate mode to do this. The development of geographical skills to create a range of simple and complex graphs and maps is important. Spatial technologies and programs such as Microsoft Excel should be used to allow students to graph a larger amount of data and experiment with different representations.

Examples of pedagogy associated with the teaching of graphic construction can be found on GeogSpace within the Inquiry and Skills illustrations of practice. A variety of geography skills books are also available for teachers, providing step-by-step instructions on the creation of the graphs and maps mentioned in this chapter.

Pause and think

Teachers often provide students with a geographic data set they have selected for a purpose, and tell them which graphic to create in a lesson involving graphicacy. How could a lesson like this be changed so students make decisions and co-construct their own learning?

Decoding – strategies for interpreting graphical data

Considering that graphicacy is a literacy, teachers should approach graphical literacy in the same way that they approach the teaching of reading comprehension. The following questions could be asked when interpreting a map, graph, diagram or image:

- What is the main idea?

- What details support the main idea?

- What is the author's purpose?

- How are the details interrelated?

- What new vocabulary or new symbols are used?

Teachers should draw attention to graphics before, during and after reading comprehension tasks every lesson, and not only when students are being asked to encode or decode graphic information. Students also need to have a strong understanding of the relevant content so they can apply what is being learnt in meaningful reading situations (Gillespie 1993).

As with written texts, students should be encouraged to be critical consumers of maps and graphs. Geography students should be provided with opportunities to question the accuracy, perspective and bias of these tools. A lesson that looks at misleading graphs and maps can strengthen a student's graphicacy, and would engage students to question graphical text in their everyday lives.

The higher order cognitive processes of interrogating, analysing and predicting graphics are challenging skills for students to develop. These cognitive processes require strategies such as teacher modelling, think-alouds and the use of examples to support student learning. The additional web materials for this chapter provide examples of teaching practice that would support students to interrogate and analyse graphs.

Short-answer questions

1. What is the difference between decoding and encoding graphic information?

2. Why do students find it challenging to represent geographic data?

3. Why would interpreting graphs or maps be similar to the steps taken to decode texts when reading?

Conclusion

Graphicacy is an essential skill for the twenty-first century learner. For teachers in the geography classroom, technology has provided unlimited opportunities to further develop graphical literacy. With this improved technology comes an increase in access to geographical data and information. Geography teachers should continue to ensure that graphicacy plays an essential role in the geography toolbox for all students.

Bringing it together

1. What are the key elements that contribute to graphicacy?

2. How prevalent is graphicacy in the Australian Curriculum: Geography? Which particular strand of the curriculum develops graphicacy?

3. Provide one example of each of the types of graphicacy that would be part of a Year 7 Geography unit on Water in the World.

4. Identify a lesson where you could apply a scaffold, thinking strategy or the use of modelling or examples when teaching about graphicacy. Consider how this strategy could be used with students to co-construct their learning.

5. Develop a series of lessons in a unit of work that you will be teaching as part of your course requirements. In these lessons, include the explicit instruction of one of the types of graphicacy that is relevant to the topic you will be teaching. Consider both the encoding and decoding of this graphic in the lesson series.

References

Australian Bureau of Statistics [ABS] (2018) *Population projections, Australia*, Canberra: ABS, retrieved from www.abs.gov.au/statistics/people/population/population-projections-australia/latest-release

Australian Curriculum, Assessment and Reporting Authority [ACARA] (2020a) *F–10 Curriculum: General Capabilities – Literacy*, Canberra: ACARA, retrieved from www.australiancurriculum.edu.au/f-10-curriculum/general-capabilities/literacy

——(2020b) *F–10 Curriculum: Geography – Structure*, Canberra: ACARA, retrieved from www.australiancurriculum.edu.au/f-10-curriculum/humanities-and-social-sciences/geography/structure/

——(2020c) *F–10 Curriculum: Humanities and Social Sciences*, Canberra: ACARA, retrieved from https://australiancurriculum.edu.au/f-10-curriculum/humanities-and-social-sciences/hass

——(2020d) *F–10 Curriculum: Geography*, Canberra: ACARA, retrieved from https://australiancurriculum.edu.au/f-10-curriculum/humanities-and-social-sciences/geography

——(2020e) *F–10 Curriculum: Geography – Rationale*, Canberra: ACARA, retrieved from https://australiancurriculum.edu.au/f-10-curriculum/humanities-and-social-sciences/geography/rationale

——(2020f) *F–10 Curriculum: Geography – Glossary*, Canberra: ACARA, retrieved from https://australiancurriculum.edu.au/f-10-curriculum/humanities-and-social-sciences/geography/glossary

Balchin, W.G.V. (1972) Graphicacy, *Geography*, 57, 185–95.

Balchin, W.G.V. & Coleman, A.M. (1965) Graphicacy should be the fourth ace in the pack. *The Times Educational Supplement*, 5 November.

Bureau of Meteorology [BoM] (2021) *Twelve-monthly rainfall deciles for Australia 1 January to 31 December 2020*, Canberra: BoM, retrieved from www.bom.gov.au/climate/maps/rainfall/?variable=rainfall&map=decile&period=12month&region=nat&month=12&day=31&year=2020

Danos, X. (2014) *Graphicacy and culture*, Loughborough: Loughborough Design Press.

De Jager, A.E. (2014) The importance of visual literacy for a changing geography, in N. Valanides, ed., *Visual literacy in the 21st century: Reconceptualising visual literacy: Describing reality, creating imagery and deciphering visuals*, n.p., pp. 93–104, retrieved from www.researchgate.net/profile/Anna-De-Jager/publication/305408651_The_importance_of_visual_literacy_for_a_changing_Geography/links/57c05dbc08ae2f5e b3321ee0/The-importance-of-visual-literacy-for-a-changing-Geography.pdf

Fry, E. (1981) Graphical literacy, *Journal of Reading*, 24(5), 383–9.

Gillespie, C.S. (1993) Reading graphic displays: What teachers should know, *Journal of Reading*, 36(5), 350–4.

Hall, M. & Russac, P. (2011) Designing information: The need for graphicacy, *The ASIDE blog – Innovation Design in Education*, retrieved from http://theasideblog.blogspot.com/2011/07/designing-information-need-for.html

Kleeman, G. (2017) *Geography literacy unlocked*, Sydney: Australian Geography Teachers Association.

Marzano, R. & Kendall, J. (2007) *The new taxonomy of educational objectives*, Thousand Oaks, CA: Corwin Press.

Poracsky, J., Young, E. & Patton, J.P. (1999) The emergence of graphicacy, *The Journal of General Education*, 48(2), 103–10.

Roux, C. (2009) Enhancing learning and comprehension through strengthening visual literacy, *Per Linguam*, 2(2), 46–60.

Tufte, E. (2001) *The visual display of quantitative information*, 2nd ed., Cheshire, CT: Graphics Press.

Wilmot, P.D. (1999) Graphicacy as a form of communication, *South African Geographical Journal*, 81(2), 91–5.

Working with data

Rebecca Nicholas

Learning objectives

By the end of this chapter, you should be able to:

- identify the different types of geographic data
- consider how geographic data is collected and represented in the geography classroom
- explore the role of data visualisations and infographics in presenting geographical data
- identify strategies to support the explicit teaching of skills associated with data in the geography classroom

Introduction

Over the last 30 years, information in the form of digital data has become the foundation on which decisions are based. Governments, businesses and organisations have realised that they must have access to the right data at the right time, and also undertake various types of analysis to make correct decisions. Data have become a part of everyday life, and the capacity for all citizens to have some level of data literacy is becoming increasingly important. Data literacy in schools, or the ability to understand and use data effectively to inform decisions, will create transferable skills for students moving into the twenty-first century workforce. This means students need to be highly skilled in collecting, recording, accessing and representing data. They should know where and how to access data, and how to use data to tell their story. Geography plays an essential role in the school curriculum to develop these skills, particularly in secondary school. While 'using data' is a skill in many subjects, geography provides the opportunity for data to be used in real-world contexts, and the skills are very transferable to many future pathways.

Data in geography

What are data?

The Australian Bureau of Statistics (2019) identifies that data 'are measurements or observations that are collected as a source of information'. In the geography classroom, what data look like varies across the curriculum. Data could include statistics or population data, data collected by conducting a survey, or data counts conducted in the field.

There are a number of differing terms associated with data, including data items, data observations and datasets. It is important for students to understand these terms so they are able to navigate the different types of data when they are collecting, analysing and representing them.

What are the different types of data?

Data can be classified as either primary or secondary. Primary data are collected first hand or may be unprocessed raw data. Most primary data used in the geography classroom are collected during fieldwork. For example, data collected as part of a vegetation transect are primary data. Secondary data are gathered from secondhand sources – which means they have already processed the data to some degree. The use of satellite imagery to study landcover change at a local scale is an example of secondary data. These secondary data would be used to support data interpretation and analysis of primary data collected in the field. Compared with other scientific disciplines, both human and physical geographers use larger amounts of secondary data (Montello 2012, p. 36). This is because geographers often study phenomena at large spatial and temporal scales.

Within these classifications, two main types of data are collected. Data can either be quantitative or qualitative. *Quantitative data* are expressed in numerical form and most often provide a count or measure. Quantitative data generally answer questions such as How much?, How many? or How often? An example of quantitative data would be the number of cars passing a point over one minute when conducting a traffic survey in the field, or the temperature and humidity readings at a location in the field. *Qualitative data* describe information. They are considered categorical data as they are not always measured in numbers, but rather describe qualities or characteristics. Qualitative data are collected using questionnaires, observations and interviews. An example in the geography classroom would be the collection of data in a liveability survey.

In geography, geospatial data play an important role. Geospatial data are all about objects, or the phenomena that have a particular location on the surface of the Earth. The advancements in the use and access of global positioning systems (GPS) now mean that any data collected can be georeferenced to a location. This is significant for geography students in the classroom as it means they can now use mobile devices to collect both quantitative and qualitative data in the field, which can be visualised and analysed in real time using applications such as geographic information systems (GIS). Geospatial data are also readily available on various websites for students to download and transform as part of a geographical inquiry.

Pause and think

How have innovations in spatial technologies changed the ways in which data can be collected and analysed? Are these same changes evident in the geography classroom?

Data in the geography classroom

Data form a large section of the Geographical Inquiry and Skills strand in the Australian Curriculum: Geography. Within this strand, students are:

- collecting, recording, evaluating and representing geographical data
- interpreting and analysing geographical data
- drawing conclusions based on the analysis of the data
- communicating and presenting findings in a range of forms selected to suit a particular audience and purpose (ACARA 2020).

The development of a student's skill in using geographical data is significant. Not only is this important in terms of graphicacy (as mentioned in Chapter 3), but also the ability to be a discerning researcher and collector of primary and secondary data. In the twenty-first century, these are the skills that will develop 'global citizens', highlighted in the rationale and aims of the Australian Curriculum: Geography.

Collecting and recording primary and secondary data

In the geography classroom, students will collect and record data when completing fieldwork, undertaking a research inquiry or completing lesson activities. All fieldwork requires the collection of some sort of data. This includes visual data (e.g. field sketches and observations) or data that can be recorded (e.g. survey results or water quality readings). Data traditionally were collected using field booklets, including field sketches sketched by hand, and quantitative and qualitative data were written down. Online and mobile-based tools are now available that allow students to record data digitally while in the field. (Please refer to Chapter 9 on fieldwork, which provides further detail on the different types of data collection.)

Students are also required to collect and record secondary data when conducting a geographic inquiry. For example, students in Year 9 geography may complete a research inquiry task that investigates the interconnections involved in the creation and sale of a product, such as a mobile phone. As part of this investigation, students will research and use secondary data from websites such as that of the World Bank (www.worldbank.org) to create graphs, maps and diagrams that support their findings. Teachers may develop a research inquiry template that supports students in the research process, particularly scaffolding the process of selecting raw secondary data and deciding how best to represent these to support the inquiry question.

In both these examples, students will need explicit instruction to model the decision-making processes involved when deciding what data should be collected. Students generally find understanding data and their significance quite challenging. Teachers should not make assumptions regarding a student's understanding of the types of data they need to collect and, most importantly, why they need to collect them. Time should be spent developing skills, particularly in junior secondary geography, so students are able to collect, record, evaluate and represent data successfully in both fieldwork and when completing a geographic inquiry.

Teachers should provide learning activities that ask students to question why data are collected, and how they can be used to make decisions.

Pause and think

What would the process of data collection look like in either a fieldwork task or a geographic inquiry? Using the Australian Curriculum: Geography, develop a step-by-step process to show the collection of data for a specific year level.

Representing data

By Year 10 geography, students should be able to represent multi-variable data in a range of appropriate forms, both with and without the use of digital and spatial technologies (ACHGS074) (ACARA 2020). The cognitive demands for students to find or collect data, 'clean the data', then determine the best way to represent these data are challenging. Cleaning the data refers to synthesising a raw data table by deleting extra columns of data that are not needed, as well as ensuring all columns and rows of data are evident.

Students are required to determine which is the best graph, map or table (either simple or complex) to represent the data in a way that allows the reader to clearly see patterns and trends, and determine next steps. This decision-making process is complex, particularly as each decision can affect the meaning being derived from the data. The development of this skill should be considered when mapping graphicacy across units, both vertically and horizontally, from Years 7 to 10.

The Connection box provides information on a sample learning experience to support students when making decisions about how best to represent data.

CONNECTION

The challenge of choosing and representing data

The following lesson activity could be used in a junior secondary or senior classroom. The activity has been designed to support students in the development of their skills in 'representing' and 'transforming' data insightfully. The activity should be conducted in groups.

- **Step 1:** The teacher provides student groups with two tables of data. The data should relate to the unit of work currently being studied. For example, this activity would work well at the beginning of the Geographies of Human Wellbeing unit in Year 10, or the Place and Liveability unit in Year 7. The quantity and complexity of data would vary depending on the year level and the time of year when the task is being completed.

- **Step 2:** The teacher provides students with a research question – for example, 'Do countries with a higher Gross Domestic Product (GDP) have better access to health?' or 'Does a higher percentage of car ownership in a suburb lead to greater traffic congestion?'

- **Step 3:** In groups of four, students decide how best to represent the data to address the inquiry question. Students use technology to transform the data and follow graphic or cartographic conventions.

- **Step 4:** Each group shares its data representation with the class. The teacher could ask each group to present or complete a 'gallery walk' to view the data representations. There is an option here for students to provide feedback on the differing representations. This could be done by asking students to write on two Post-it Notes – a star for what they liked about the representation, and a question. The groups can then use this as peer feedback to improve their data representations.

- **Step 5:** The teacher facilitates a discussion: 'Which data representation is the most effective in addressing the inquiry question? Why?'

- **Step 6:** Through the discussion, the teacher and class develop a checklist to support students when deciding how best to represent data.

This lesson could be repeated throughout a year or stage, and would be useful to complete before undertaking fieldwork.

Questions

1. What challenges could the teacher experience when completing this learning experience with a class?

2. What sections of the learning experience would students find difficult? How could this be managed to support students in completing the activities?

3. What further modifications could be made to this learning experience to further develop or consolidate student skills in representing data?

Short-answer questions

1. What is the difference between primary and secondary data sources, as well as qualitative and quantitative data?

2. What role has spatial technologies played in improving access to data collection in fieldwork?

3. According to the Australian Curriculum: Geography, what are the differing geographical skills required by students when working with data?

Data visualisations and infographics

Data visualisation is an overarching term that refers to any instance of data being presented in a visual way. It includes the use of visual elements such as graphs and maps to provide an accessible way to see and understand trends, outliers and patterns in data. Data visualisations are readily accessible and should be used by both students and teachers in the Geography classroom.

Gapminder (gapminder.org) is a popular data visualisation that supports students to explore indicators of development. Hans Rosling created the visualisation tool in 2005 to dismantle misconceptions and promote a fact-based world-view using data. Gapminder also provides a number of video explanations of data, along with teaching resources and lesson plans to support students in their understanding and use of development indicators.

Earth.nullschool (earth.nullschool.net) is another excellent example of a data visualisation, with a focus on global weather conditions. The visualisation provides students with access to real-time weather data. Students are able to turn layers on and off to look at global or regional weather patterns. There are over 20 layers of data that include ocean currents, sea surface temperature, ocean waves, wind, aerosol pollutants, precipitation and humidity.

Data visualisations support student engagement in the classroom and provide a real-world context for geography learning. Geography teachers should ensure that unit and lesson plans include visualisations such as these to develop a deep understanding of how large amounts of data can be used to make decisions at various scales.

Infographics are similar to data visualisations as they present data visually. However, infographics are developed with the purpose of effectively communicating a narrative through the use of graphic design and text. They are widely used as they offer the author a way to take an issue or concept that is complex and communicate data so they can be understood easily (see Figure 4.1).

To further develop a student's understanding of data, infographics could be used as a tool to 'tell a story' about a geographic challenge or concept that is being studied. There are a range of free online tools to create infographics. Teachers should ensure that students have the time and space, as well as scaffolding, to create an infographic as a class activity or assessment task.

CONNECTION

Guidelines for a good infographic

The process to plan, design and create an infographic is a complex one. The following steps could be used to guide students through this process (based on Infogram 2017).

1. **What is the story?** The key to a good infographic is to draw an audience in with a captivating story or message. The story should:

    — state the problem or issue

    — provide a timeframe

    — deliver a message

    — present a solution.

2. **Data.** The data gathered for an infographic should be accurate and represent the most recent statistics available. The data need to be 'cleaned' and sorted, as well as correctly referenced.

3. **Good headline.** The headline used for the infographic is crucial. It must be short, clear and relate to the story the data are telling.

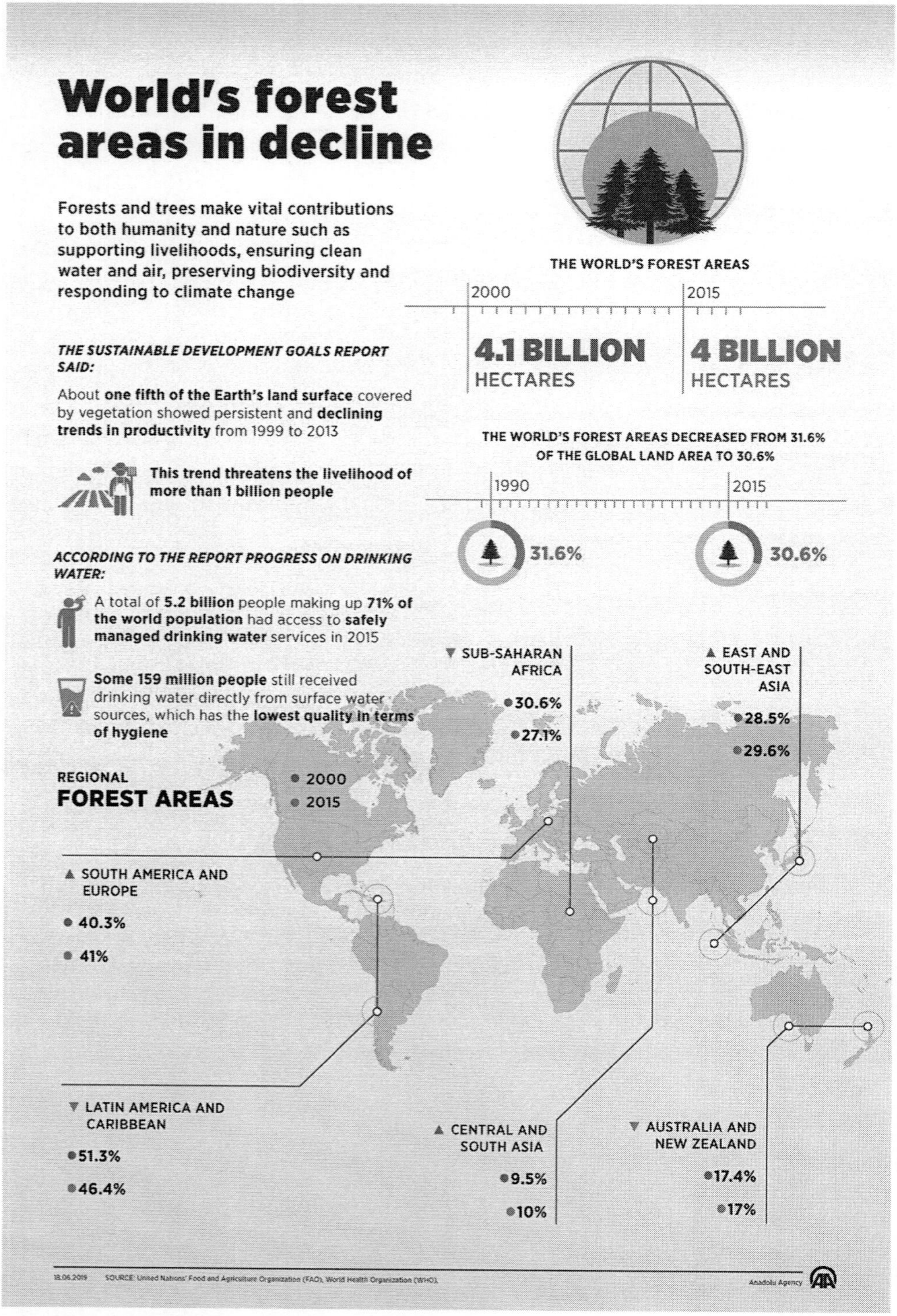

World's forest areas in decline

Forests and trees make vital contributions to both humanity and nature such as supporting livelihoods, ensuring clean water and air, preserving biodiversity and responding to climate change

THE SUSTAINABLE DEVELOPMENT GOALS REPORT SAID:

About **one fifth of the Earth's land surface** covered by vegetation showed persistent and **declining trends in productivity** from 1999 to 2013

This trend threatens the livelihood of more than 1 billion people

ACCORDING TO THE REPORT PROGRESS ON DRINKING WATER:

A total of **5.2 billion** people making up **71% of the world population** had access to **safely managed drinking water** services in 2015

Some **159 million people** still received drinking water directly from surface water sources, which has the **lowest quality in terms of hygiene**

REGIONAL FOREST AREAS
- 2000
- 2015

THE WORLD'S FOREST AREAS

2000	2015
4.1 BILLION HECTARES	**4 BILLION** HECTARES

THE WORLD'S FOREST AREAS DECREASED FROM 31.6% OF THE GLOBAL LAND AREA TO 30.6%

1990	2015
31.6%	30.6%

▼ SUB-SAHARAN AFRICA
- 30.6%
- 27.1%

▲ EAST AND SOUTH-EAST ASIA
- 28.5%
- 29.6%

▲ SOUTH AMERICA AND EUROPE
- 40.3%
- 41%

▼ LATIN AMERICA AND CARIBBEAN
- 51.3%
- 46.4%

▲ CENTRAL AND SOUTH ASIA
- 9.5%
- 10%

▼ AUSTRALIA AND NEW ZEALAND
- 17.4%
- 17%

18.06.2019　　SOURCE: United Nations' Food and Agriculture Organization (FAO), World Health Organization (WHO).

Anadolu Agency

Figure 4.1 Example of an infographic

4. **Type of infographic.** The type of infographic needs to be consistent with the original story. Types of infographics include:

   — *data centric* – showing statistical information in a variety of graphs, tables and diagrams

   — *timeline* – showing information over a chronological period

   — *how to* – showing a step-by step process or the progression of information

   — *geographical* – location-based infographic using maps for geographical data

   — *comparison* – comparing and contrasting two different subjects or topics

   — *hierarchical* – showing a chart with pre-defined levels or the hierarchy of a topic

   — *flowchart* – starting at a single point, then showing how the topic branches or grows

   — *list* – showing mostly text and icons, listing information about a given subject

   — *anatomical* – breaking down the subject's composition, or showing how it works

   — *photo-based* – based on a photo, using text and data to explain a point (Infogram 2017).

5. **Design and aesthetic.** The design and aesthetic of the infographic involves the creative process to determine colour, theme and structure, fonts, graphics and appearance. Colour choice, the use of colour to highlight important information, types of fonts and alignment are important design features that need to be considered.

6. **Graphicacy.** The correct types of graphs and maps need to be used when visualising data in an infographic. The four main types of graphics are:

   — *comparison* – used to compare one or more datasets; they can compare items or show differences over time

   — *relationship* – used to a show a connection or correlation between two or more variables

   — *composition* – used to display parts of a whole and change over time

   — *distribution* – used to show how variables are distributed over time, helping to identify outliers and trends (Infogram 2017).

7. **Focus on important data.** Make sure that the most compelling statistic is obvious and determine what supporting data should follow.

8. **Keep it simple.** A common mistake made when designing infographics is including too much information in one single infographic. Unless you know how to design complex data visualisations, keep it simple. Don't use too much text or too many numbers. Avoid grid lines or legends.

Questions

1. Consider a unit of work in Years 7 to 9 geography. Which of these units could include an assessment task that requires students to create an infographic? Why would an infographic be a suitable task for this unit?

2. What skills would you be assessing if students were required to design and create an infographic as part of an assessment task? Use the Year 7 or 8 Achievement Standards in the Australian Curriculum: Geography document and highlight these.

Short-answer questions

1. What is the difference between a data visualisation and an infographic?

2. How would data visualisations improve student engagement and understanding of data in the geography classroom?

3. What are the guidelines for a good infographic?

Engaging students with geographic data in the classroom

Geography students are frequently required to interpret and analyse data that include very big numbers. Students often do not comprehend the size and scale of their own settlement, and therefore can find the raw number associated with data difficult to understand. For example, students analysing global data such as country size or GDP find it difficult to comprehend what these data actually mean. An understanding of geographical scale is vital to any piece of spatial analysis.

These concepts are also challenging for geography teachers, as it is difficult to find learning experiences that will illustrate the relationships between them.

CONNECTION

The importance of geographical scale and data

The following lesson idea asks students to consider the role of geographical scale when looking at data.

Questions

Complete the lesson below prior to teaching it with a class, and answer the following questions:

1. What did you find challenging when you completed this activity?

2. Considering this, how could the lesson be modified to support students to understand these concepts following a lesson sequence?

3. What do the students need to know and do prior to this lesson idea?

4. What would the next lesson look like in this sequence to ensure all students have understood the concepts?

5. In what year level and unit could this lesson sequence be found in the Australian Curriculum: Geography for Years 7 to 12? Why?

Lesson idea

Patterns revealed by census data vary depending on the scale at which the data are viewed. The more local the level, the more robust the data and evidence. The following activity encourages students to identify and understand Australian Bureau of Statistics data through the lens of differing scales. Students will look specifically at age structure, moving from national and state data, down to local government area (LGA), then finally focus on Statistical Area 1 (SA1) data, which represent the smallest unit of release from the ABS. SA1 data include 200–800 people in the geographic area.

Activity

- **Step 1:** Using the Australian Bureau of Statistics website (www.abs.gov.au), find data and create a national map that shows the change in the number of 0–4-year-olds between 2011 and 2016. Students should interpret and analyse the pattern shown on the map.

- **Step 2:** Select a capital city that shows the highest growth in this age group (Melbourne or Sydney). Interpret and analyse the pattern shown for local government areas (LGAs) in one of these cities.

- **Step 3:** Identify significant areas of growth in this age group by looking at Statistical Area 3 (SA3) data. Focus further and interpret any patterns using SA2 and SA1 data.

- **Step 4:** Ask students to select another data set to help them explain why this is an area of growth (e.g. dwelling structure). Students may also need to complete some research on the land use of the suburb to see whether this has changed over time.

- **Step 5:** Students should share their findings with a partner.

- **Step 6:** Students should use their observations to write a paragraph exploring the following statement: 'Geographic scale plays a significant role when analysing data for patterns, trends and relationships.'

Pause and think

According to the Australian Curriculum: Geography, when do students begin collecting different types of data as part of their classroom or assessment tasks? How could the activity above be differentiated for a Year 7 geography class? Or a Year 12 geography class?

Conclusion

Working with data is the core business of geographers. Geography students should be exposed to data regularly, either in a raw and rough state in an Excel spreadsheet, or as an engaging data visualisation. Developing the skills of understanding what data are, how and why they are being collected and recorded, and then how to represent them to tell a story are essential skills. Innovation and changes in technology have meant that data sets are readily accessible for both students and teachers.

Bringing it together

1. Identify two examples of primary data, secondary data, qualitative data and quantitative data that would be used in the geography classroom.

2. What geographical skills are associated with data in the geography classroom?

3. How can geospatial technologies be used in the classroom to develop students' skills in interpreting and analysing data?

4. Why would students find representing and transforming data challenging? How could students be supported in the development of these skills in the classroom?

5. Create a two-lesson sequence prior to fieldwork or the start of a geographic inquiry that focuses on the skills required to collect and record data. What teaching strategies could be included to ensure that students understand why and how to collect data?

6. Develop an assessment task that requires a junior secondary student (Years 7–10) to create an infographic associated with one of the units. The task should include both a checklist for creating the infographic and a rubric or marking guide to assess their work.

References

Australian Bureau of Statistics [ABS] (2019) *Statistical language – what are data?*, Canberra: ABS, retrieved from www.abs.gov.au/websitedbs/D3310114.nsf/Home/Statistical +Language+-+what+are+data

Australian Curriculum, Assessment and Reporting Authority [ACARA] (2020) *F–10 Curriculum: Geography*, Canberra: ACARA, retrieved from https://australiancurriculum.edu.au/f-10-curriculum/humanities-and-social-sciences/geography

Infogram (2017) How to make an infographic in 10 steps, retrieved from https://infogram.com/blog/make-infographic-10-step-guide

Montello, D. (2012) *An introduction to scientific research methods in geography and environmental studies*, Thousand Oaks, CA: Sage.

CHAPTER
5

Fieldwork skills

Stephen Cranby

Learning objectives

By the end of this chapter, you will be able to:

- identify the contribution of fieldwork to the development of student skills
- recognise the importance of student skill development in fieldwork planning
- distinguish between generic 'transferable' skills, geographical thinking skills and specific geographical fieldwork skills
- distinguish between fieldwork skills, techniques and applications, and the tools that can be used to gather geographic fieldwork data

Introduction

Looking at the world 'through the eyes of a geographer' finds its deepest expression through the practice of fieldwork. Fieldwork brings together the ideas of 'tropophilia' (Anderson & Erskine 2014) – the need for a person to experience, and to be engaged and challenged in, their sense of place – with that of 'topophilia' (Tuan 1974), which emphasises the need for a person to develop 'positive connections to places'. It is in the field, through the application of geographical skills, tools and techniques, that students make connections between the real world and classroom learning.

Our increasing awareness of the environmental impact of human activities has broadened the scope of geographical studies. In response, fieldwork has developed a broad range of skills, tools, and techniques for gathering fieldwork data. With the digital age, the availability of new tools and techniques expands the possibilities for data collection in the field, and requires new skills to be developed (Esteves, Hortas & Mendes 2018).

This chapter examines skills developed by, and brought to play in, fieldwork. Progressing from generic skills used and refined through fieldwork, the discussion focuses on the geographical nature of skills used across all fieldwork activities, and looks at the key geographical skills and tools that can be drawn upon to construct authentic fieldwork experiences for students.

Fieldwork has always been an important facet of geography, helping to inform, validate and consolidate the study of people and place. To this day, fieldwork remains rather simple and straightforward, involving the gathering of primary data in the field. The 'process' of fieldwork occurs through the use and application of a wide variety of geographic and generic skills. The following discussion of fieldwork skills will examine the place of:

- fieldwork skills in students' wider learning

- fieldwork skills for thinking geographically

- specific geographic fieldwork skills

- geographic fieldwork tools and technology.

Fieldwork skills in students' wider learning

Geography teachers focus on the geographic skills needed to encourage learning through students' fieldwork experiences. But what of the 'transferable' and 'social' skills that develop because of students' fieldwork experiences? What part do they, or should they, play in a student's fieldwork experience? The following will identify and address these questions through the examination of:

- generic and transferable skills

- fieldwork design in skill development

- fieldwork's development of cognitive skills

- sequential development of skills.

Generic and transferable skills

Many studies acknowledge that fieldwork provides greater opportunities for broader skill development than just geographic skills. These general skills 'often get overlooked as a benefit in the drive for subject knowledge' (Wheeler et al. 2011, p. 19). So what potential skill development is overlooked? Fieldwork provides many opportunities to develop and enhance independent learning and thinking skills. These skills and others (e.g. study, personal and life skills) are highly transferable across the curriculum and life more generally.

These broader skills can loosely be grouped into what may be described as 'basic' and 'higher level' skills. When looking at these in the context of their acquisition within a fieldwork exercise, they can further be identified as either 'direct – those skills that are explicitly addressed during the fieldwork activities, or, indirect – those skills that are acquired implicitly through fieldwork experience' (Wheeler et al. 2011, p. 17). Table 5.1 organises this diverse list of skills to identify their place in fieldwork.

Table 5.1 Fieldwork's generic and transferable skills

Direct		Indirect	
Basic skills	**Higher-order skills**	**Basic skills**	**Higher-order skills**
Working independently	Independent learning and organisation	Autonomy and personal organisation	Autonomous learning and systematic organisation
Group work	Teamwork	Collaborative work and interpersonal	Leadership and negotiating
Communication		Listening	Open to new ideas
Observation			
Collecting and recording data	Making notes	Summarising information and time-management	Organising ideas coherently
Deciding	Problem-solving and reasoning		Innovation
Choosing	Deducing		Applying logical thinking and reflection
Speaking and questioning	Conducting interviews		
Using diagrams and sketches	Geographical representation		Creativity
Gathering numerical data	Manipulating numerical data and calculation	Numeracy	
Using IT	Computer literacy		

Source: Based on ideas presented in Foskett (2000); Nowicki (1999); Rydant et al. (2010); Wall & Speake (2012); Wheeler et al. (2011).

Development of any of these skills during fieldwork will depend on the teacher's objectives and the fieldwork's structure. Table 5.1 provides a mechanism to identify and consider what opportunities exist to develop these generic skills in any fieldwork exercise. As Woolhouse (2016, p. 45) notes, 'fieldwork provides an ideal platform to push students to higher order thinking and the experiential learning aspect of fieldwork actually accelerates the learning of these thinking skills'. Fieldwork would benefit from identifying and acknowledging its contribution to enhancing these generic skills as well as its subject-specific skills.

Fieldwork design

The choice and design of fieldwork activities and learning outcomes can influence the development of students' skills. This has implications both for planning fieldwork, and for re-evaluating current practice (Peasland et al. 2019; Wall & Speake 2012). The choice of teaching-learning methodology – for example, the balance between teacher-led and student-initiated activities – can inhibit or promote opportunities for meaningful learning and skill development.

Designing fieldwork around real-world contexts and problems invites greater ownership and involvement from students. Combined with structured inquiry-based or problem-solving approaches, students are encouraged to find solutions to something that affects their everyday life. Increased purposeful involvement leads to greater attention to tasks undertaken, and consequential skill development. Fieldwork site selection is pivotal in providing opportunities for the development and use of a wide range of skills. Thought must also be put into opportunities to develop wider skills to complement and reinforce the geographic skills undertaken (Esteves, Hortas & Mendes 2018; Foskett 2000).

Fieldwork's development of cognitive skills

By its nature, fieldwork stimulates and challenges students into new ways of working and cognitive thinking. The process can also lead to growth of the affective domain, which research has shown can lead to improvements in cognitive learning. In designing fieldwork tasks and investigations, it is important to select appropriate techniques and skills that draw on higher order thinking to enable deep learning to take place (Foskett 2000; Wheeler et al. 2011; Woolhouse 2016).

Thinking skills and processes more commonly addressed in fieldwork fall into the knowledge and performance levels of Figure 5.1. Opportunities for extending students' skills through the metacomponent level occur more frequently when built into the fieldwork structure. Transfer skills (lateral and vertical) less often addressed in fieldwork contribute to students' higher order thinking skills more often during inquiry learning or problem-solving exercises.

Bloom's revised taxonomy displays four knowledge dimensions interacting with six cognitive processes (see Figure 5.2). These dimensions and processes range along a continuum of thinking skills from lower to higher order. The shaded areas indicate

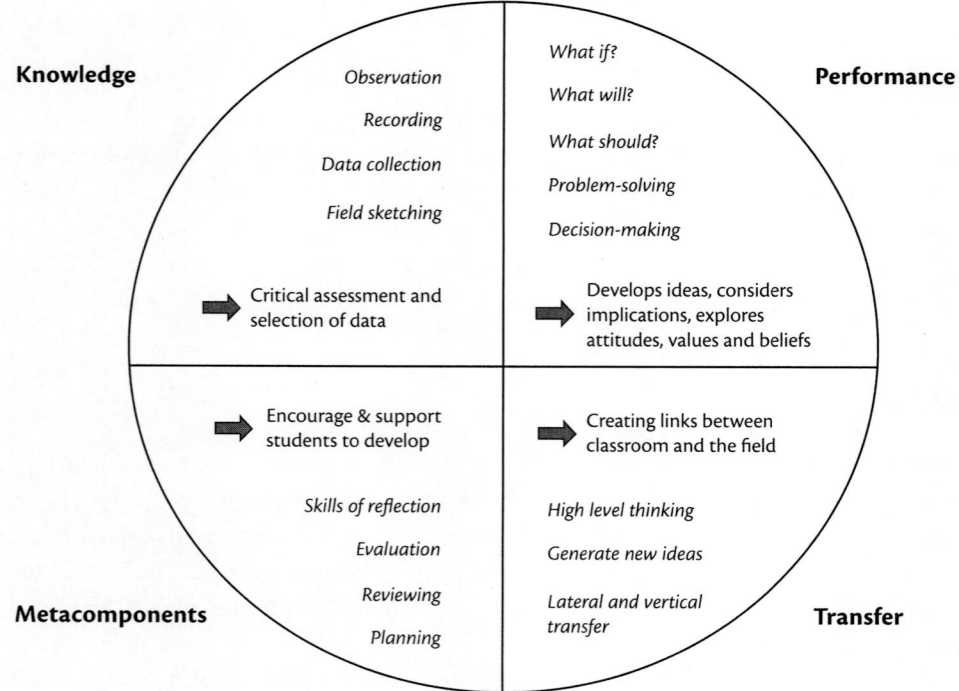

Figure 5.1 Thinking skills

Source: based on ideas presented in Foskett (2000).

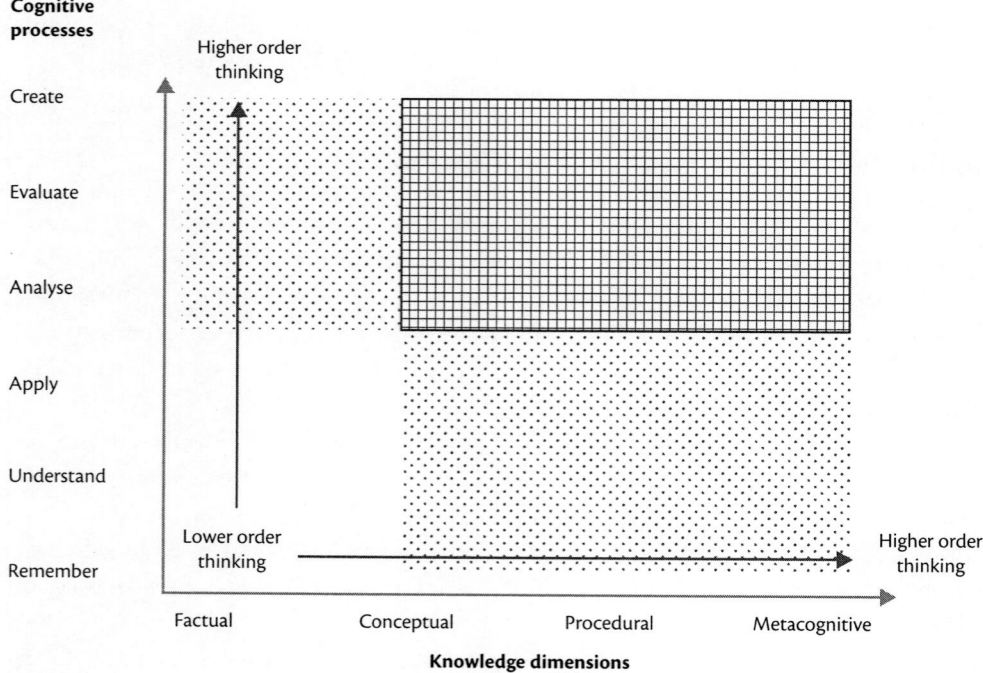

Figure 5.2 Bloom's taxonomy and thinking skills

Source: based on ideas presented in Whalen & Paez (2021) and Anderson et al. (2001).

processes and dimensions that provide for more challenging and higher-order thinking. It is important to construct fieldwork tasks that engage students in these areas, in order to maximise these opportunities (Whalen & Paez 2021; Anderson et al. 2001).

Sequential development of skills

Fieldwork skills lend themselves to being introduced progressively, both in complexity and depth. The earlier students use a skill, the more familiar and competent they will become. Over time, students develop a 'bank of skills' to be drawn upon (Field 2009, p. 44). Students' introduction to new fieldwork skills depends on the context and subject matter of each exercise, their age level and general skill development. The same skill can be introduced in the early years and developed to a higher order in the later years.

This sequential development of skills needs to be planned and integrated into a school's fieldwork program. There is no one progression formula. Research examining the sequentiality of fieldwork skills poses several questions as a reflective tool for teachers to evaluate skill progression within their fieldwork program:

- Are the fieldwork skills repeated, are they repeated with greater complexity or are new skills developed in each year?

- Is the development of skills evident to the teachers and students?

- Is the increase in complexity over time focused on data collection, analysis or interpretation?

- Does the fieldwork draw upon skills gained in previous years? (Woolhouse 2016, p. 40)

Short-answer questions

1. Write a brief explanation of fieldwork's contribution to students' wider learning for a Year 10 student information night brochure.

2. Of the areas noted, which are the hardest or easiest to build into a fieldwork task? Why?

3. Which of the areas demands the most of you as a teacher? In what way?

Fieldwork skills for thinking geographically

Fieldwork skills should encourage or facilitate students to think geographically. Teachers should select the skills for each fieldwork task to maximise students' growth and development geographically (Esteves, Hortas & Mendes 2018). Several fieldwork skills can be described as 'universal' and applied to any fieldwork exercise, be it physical or human focused, or rural or urban located. When revisited over successive years and fieldwork exercises, these skills play an important part in developing students' ability to think geographically. They include:

- observation
- note-taking/recording
- sampling.

Observation

Fieldwork is about data collection in the field. The first form of data collection, and the key to any fieldwork exercise, is observation. Appropriate and effective observation skills are the stepping stone to higher level skills. Research indicates that, among other forms of data collection, observation is the most memorable to students (Wall & Speake 2012; Wheeler at al. 2011).

It is important to create opportunities for students to record/note observations and interpretations during their fieldwork. The provision of appropriate techniques – for example, structured worksheets – assists students through the observation process. Students gathering data through observation is crucial to them creating knowledge. Observations provide the foundation to question and seek explanation for the features, patterns or processes occurring. They provide relevance and context for the use of other geographical skills and techniques to seek answers to such questions (Wheeler at al. 2011).

As students experience sequential fieldwork exercises, the scaffolding provided to make and record observations becomes more sophisticated, and their observation skills develop. Though students may initially not fully appreciate the skill of observation as much as measuring longshore drift or throwing a tennis ball into a river to measure velocity, in time they come to learn that it is the most important skill when it comes to fieldwork (Wheeler et al. 2011).

Formal or structured observation tasks help to ensure that students are looking beyond the more immediate data-collection tasks of surveys and counts. Opportunities for informal observations take place through general discussions, while sitting around during a break or, more frequently, during overnight fieldwork trips, sitting around talking about the day's work and planning for the next day. Though unstructured and relaxed, these moments can develop insights into the fieldwork that may otherwise be missed (Wheeler et al. 2011).

Fieldwork, and specifically observation skills, embed the habit of looking at the world through a 'geographer's eyes'. No more walking through life and experiencing the world as if through a tunnel. As a student commented, 'Knowing about geography and the environment has spoiled my walks ... I'm always thinking about what I'm looking at now!' (Wheeler et al. 2011, p. 17).

Note-taking/recording

Fieldwork provides only one opportunity to observe and collect data, unless there are opportunities for multiple visits. Back in the classroom, one can only work with the data collected, as memory is fallible. The provision of ubiquitous 'worksheets' scaffolds student observations, measurements, reflections, assessments and general thoughts about the fieldwork exercise into useful notes and records.

The choice, design and degree of scaffolding of worksheets are determined by the students' age level, the subject matter and the degree of complexity required in the task. Note-taking/data-recording worksheets that may be included within a fieldwork 'booklet' can include a combination or derivative of any of the examples shown in Table 5.2.

Table 5.2 Examples of note-taking/data recording resources

Resource	Explanation
Base maps	Outline maps of fieldwork sites to record features, observations, photographs taken, etc.
Lined paper	Space for open-ended/unstructured notes and observations
Location maps	Indicating fieldwork site detail, the surrounding region, and other maps providing secondary data to inform activities
Photo logs	To record each photo's number, location, direction and purpose
Plain paper	For additional notes/observations, extra field sketches/maps or to replace mistakes made earlier
Quadrat, sketch map and transect pro-formas	Blank pro-formas for recording data in the field simply and accurately in a format useful for transcription afterwards
Structured questions	Based around activities occurring during the fieldwork, where key observations, explanations, hypotheses and reflections are drawn out
Tables and matrices	Pre-determined blank tables and matrices requiring students to complete rankings, ratings, observations, assessments, scores and counts, etc.

Relevant worksheets and other resources are generally supplied to students in a fieldwork booklet on the day. These can be supplemented with smartphone or tablet apps and other digital formats to record, collect, collate and transfer data collected in the field to students and/or class computers (see Chapter 6 for further discussion). Planning for these tools and data recorders needs to be integrated with the worksheets and other resources provided in fieldwork booklets. Photographs and videos are other ways to record observations made while on fieldwork.

Holton (2017, p. 200) observes that when students are making notes and recording observations in the field, they have a 'more productive field-learning experience' and there is a shift in 'learning from being passive to active'. There is a possible discontinuity between the public's perception of fieldwork and that of geography teachers – that some students do the work, while others reap the benefits. The opposite is true. Fieldwork is the pinnacle of cooperative learning. Sharing of ideas, observations, notes and data collected contributes to the learning of both the sharer and the receiver. The sharer restructures, clarifies and embeds their understanding of their ideas and notes as they discuss them. The listener absorbs their understanding of what is shared and re-codes it into their own perspective. Each is learning on the spot, and they are both richer for the experience – which is the real purpose of fieldwork.

Pause and think

In what way(s) can fieldwork, through its cooperative nature, provide opportunities to structure the learning around group work, team-building, leadership and negotiation?

Sampling

Although the aim of sampling is to 'make the content more manageable for both the teacher and the student' (Caton 2006, p. 61), the choice of sampling method is predominantly the responsibility of the teacher. Senior students can and should be involved in the design of the sampling method so they are aware of the reasons for its selection, and its strengths and limitations, enabling them to reflect on and evaluate the results in a more meaningful way.

Choice of sampling method is not limited to the conduct of surveys of people. It applies in a multitude of situations as diverse as selecting the size, number and distance apart of quadrats along a vegetation transect or selecting the time(s) and frequency of conducting a count of users of a local park. Sampling methodology is an integral aspect of fieldwork design. Table 5.3 describes four commonly agreed forms of sampling.

The question students often ask is, 'How many surveys do I need to complete?' Sample size needs to be large enough to make reasonable observations about the population or site studied. Too small a sample size leads to questions of validity; too

Table 5.3 Forms of sampling

Sampling method	Definition (WJEC GCSE Geography n.d., p. 3)	Example
Random	'All parts of the study area/ population have an equal chance of being selected.'	Students disperse among crowds at a Sunday market and gather three survey responses each.
Systematic	'Parts of the study area/ population are selected at equal intervals.'	A household environmental audit is conducted along the north side of each street surveyed.
Stratified	'Where the study area/ population is split into groups and then a random sample is taken within each group.'	A river is divided into 2 kilometre 'reaches', taking water samples from any accessible point within each 'reach'.
Opportunistic	'Picking sample sites or people due to their ease and convenience.'	Quadrat sites adjacent to walking track are selected rather than those further away. Similarly, people are chosen to survey based on personal judgements.

large a size can be difficult to process. Samples are selected as they are quicker and less difficult than sampling a whole population; the aim is for the sampling results to be representative of the whole site or larger population (Sharma, 2017).

The choice of sampling method is related to the fieldwork site or topic, and other variables. These may include:

- **Accessibility of site.** It is important to be able to physically collect the data, consider safety issues, etc.

- **Age and skill level of students.** Do they collect the sample individually, in small groups or collate everyone's limited data into an appropriate sample size?

- **Time.** What is the best time to collect the data? Can the site be revisited at different times?

- **Weather.** Is the planned method achievable in the expected weather conditions? Is a rethink of the methodology needed?

Short-answer questions

1. Of the three skills noted, which contributes the most to the development of students' ability to think geographically? Why?

2. As an early career teacher, what skill presents the greatest challenge for you? In what ways?

Geographic fieldwork skills

It is no wonder, given the range of environments (e.g. natural/human-made), locations (e.g. urban/rural), perspectives (e.g. sustainability/development) and other fieldwork contexts, that a plethora of 'fieldwork skills' exist in the literature. Almost 'impossible to list conclusively', they are 'therefore often referred to as the "hidden agenda" of geography fieldwork' (Wall & Speake 2012, p. 421). This complexity is amplified when a skill defined in one context can be the same skill explained in another. For example, a beach profile and urban land use profile both involve the skill of conducting a transect (see Table 5.4).

Table 5.4 Fieldwork skills

Skill	Technique	Tools	Related skills	Context
Transect	Beach profile	Tape measure, clinometer	Field sketching, recording results	Coastal
	Land use profile	Trundle wheel, sound meter, compass	Annotating base map, photographing	Urban

This situation becomes more confusing upon examination. Often, a reference to a fieldwork skill fails to distinguish between the skill described and the tool or apparatus used. Similarly, the distinction between skill and tool is blurred when discussion focuses on the technique or its application in a particular context. Sorting through this confusion, one can identify and define the 'core skills' common to fieldwork. Some exercises will draw on many of these skills, while others may only utilise a few. These skills can be found across most techniques used in fieldwork. Their characteristics include:

- **Repeatability.** For example, a traffic count or an environmental impact assessment may be different activities and contexts, but they both use the skill of making 'observations'.

- **Complementarity.** Many skills are complementary and often used in conjunction with each other. Even though the 'record sheet' and subject matter may differ in the above examples, they both use the skills of recording 'observations' on a prepared 'record sheet'.

- **Generic.** Some skills are generic, as in the application and use of 'fieldwork tools'. Even though each tool is different in purpose and usage – for example, a clinometer compared with an anemometer – the skill of its appropriate selection, effective use and recording of data output is the same.

- **Universality.** Some skills are 'universal' and must be included in the design of all geographic fieldwork. The first two are absolutes: 'direction' – the key to 'locating' themselves and site features in space; and 'observations' – the purpose of fieldwork. At least one of either reading and interpreting 'maps' or annotating 'base maps', as students must be able to locate fieldwork site(s) in relation to other local and regional features.

- **Adaptability.** All skills are adaptable and can be used in different geographic contexts, with a variety of techniques and with different tools.

Skills can be divided into 'analogue' – based on hand, eye, physical skills, pen and paper – and 'digital' – relying on the internet and other e-technologies. Analogue skills can be described as traditional, tried and trusted. Numerous examples, descriptions and explanations of their use and application exist both in print and online. Digital fieldwork skills are still at their emergent stage. Early technologies (e.g. GPS and GIS) are being incorporated into school fieldwork as accessing them becomes simpler and cheaper. Smartphone 'apps' are increasingly being incorporated into school fieldwork, as the availability and type of apps continually improve and broaden. The pedagogy for analogue skills is well recognised, whereas digital skills are still being developed and recognised.

The core skills in geographic fieldwork listed in Table 5.5 are defined and illustrated by a range of examples of techniques and uses. These examples are not exhaustive – for instance, no reference is provided to fieldwork in glaciated landscapes. Some examples may involve the use of multiple skills, though only mentioned in the context of one. Furthermore, examples or activities can illustrate more than one skill, depending on design.

Short-answer questions

1. What is the difference between a skill, technique, context and the tools used in fieldwork?

2. Are the 'skills' identified a true representation of core fieldwork skills? What might be missing?

3. What other examples of techniques, use or methods might illustrate application of the core skills identified?

Table 5.5 Core skills in geographic fieldwork

Skill	Explanation
Analogue skills	
Identifying **direction**	Recognise the compass's cardinal points, the intermediate 16 points and finer degrees. Use compass for navigation, location, and direction. *Navigate a trail using compass and map. Identify direction of photograph taken, or slope's aspect, or one feature relative to another.*
Constructing a **field sketch**	Field sketches record information as observed in the field. They illustrate general features of a view, then highlight and annotate the key features observed. *Essential tool providing a 'snapshot' record of one's field observations. Common in natural landscapes illustrating spatial relationships between features or processes taking place.*
Application and use of **fieldwork tools**	Fieldwork tools are used to measure, record and communicate relevant geographical information. The skill is in the appropriate selection of tool(s) for the task, their effective use and the recording and transfer of resultant data. *Use clinometer, and other tools to measure land characteristics (e.g. slope, aspect); other tools to test riverine (e.g. depth, velocity) and soil characteristics.*
Interpreting **images**	Use of satellite images, oblique and vertical aerial photographs in the field to locate relative positions of observed features, identify differences (features, spatial patterns) between images and field observations. *Investigate an area recovering from bushfires, using images from pre-fires and after to identify change; investigating change in built environments.*
Conducting **interviews**	Interviews have focused questions, are prepared for diversion or expansion, and result in detailed notes. Some are pre-organised, while others spontaneous. *Interviews with key stakeholders in a proposed urban redevelopment (e.g. a local councillor, government or community representative) or investigating people's lived experience of places.*
Reading and interpreting **maps**	Use six-figure grid references and scales, interpret contour patterns and use key to identify features on topographic

Table 5.5 (*cont.*)

Skill	Explanation
	maps to locate/identify/interpret locations, distances and routes. *Locate features of the fieldwork site. Navigate from one fieldwork site to another.*
Making **observations**	Noting features, people's activities, processes taking place or observer's thoughts about what they are seeing. Data recorded as notes, tallies, sketches, etc. *Complete worksheets recording observations of environmental impact, land use, pedestrian footfall, sensory mapping, traffic counts, wave analysis, etc.*
Taking **photographs**	Cameras and phones used to record images of places, physical or cultural features of the fieldwork site for later processing and analysis. *Collect photographic records in the field of coastal erosion, bushfire regrowth, urban development and change.*
Investigating **quadrats**	Survey a defined area in detail, record characteristics enabling comparison and analysis between sites in contrasting areas, or intervals along a transect. *Scale can vary from centimetres to metres for vegetation characteristics (e.g. number, type and density) to tens of metres for human characteristics (e.g. density use of a park).*
Completing **record sheets**	Use of worksheets using pre-set questions, tables, matrices, etc. enables consistent and reliable data collection. Both quantitative and qualitative data can be gathered. *Worksheets could cover building decay survey; measuring longshore drift; pollution assessment (e.g. air quality, water); residential housing survey; or weather characteristics.*
Completing a **site analysis**	Worksheet detailing a set of criteria (observations, questions, tasks – often based on curriculum requirements) about specific site(s), enabling comparison and analysis later. *Topic-specific worksheets include environmental management assessments, disabled accessibility analysis or physical environment characteristics of sites.*
Annotating **base maps**	A bird's eye view of the site enabling features to be annotated on base map, with key outlining symbols/

Table 5.5 (*cont.*)

Skill	Explanation
	colours used, and brief annotations provided. *Identify, locate, and map features of a local park, a shopping strip, a coastal area or walking track management controls.*
Drawing **sketch maps**	Essential in all fieldwork exercises. A site's significant features are drawn freehand. Appropriate scale estimated, key developed, and brief annotations provided. *Gather data on the layout of facilities in a recreation reserve or coastal features.*
Conducting **surveys**	Focus questions on a topic's key issues. Can include both open or closed questions. Sampling process, site, and purpose are considered. *May seek data on people (age, gender), location, purpose, behavioural and attitudinal characteristics, also people's perceptions about an issue, project or development (e.g. accessibility or quality of life).*
Completing a **transect** analysis	A line between two points along which observations, measurements and recordings are made. These may occur at regular intervals (in conjunction with use of quadrats) or be continuous over the transect. *Transects can gather data on beach profiles, land use change, plant succession or noise profiles.*
Making a **video**	Recording processes or features observed, interviews made or a guest speaker, in the field for later reference, note-taking and analysis. *Record processes of coastal erosion, traffic patterns, spatial distribution of commuters and field interviews.*
Digital skills	
Smartphone apps	Smartphone apps used in the field to identify, measure, count and record data for later processing in class. *Expand and develop, with many supporting spatial thinking and active data collection (e.g. ESRI Collector for ArcGIS).*
GPS	Identify and record the coordinates of points of interest or features for later, adding to a map or downloading into a GIS for processing and analysis. *Capture location data for photographs, water and soil sampling sites, etc.*

Table 5.5 (*cont.*)

Skill	Explanation
GIS	Used to input data in the field or to ground truth secondary data. *Import field data in real time to identify gaps and capture the missing data while still in the field.*

Note: Although the internet, virtual reality (VR) applications, Google world/stories/expeditions, and other developing digital technologies require students to develop skills in their use and application, they do not qualify as 'fieldwork skills' as they lack the prerequisite collection of 'primary data in the field'.

Geographic fieldwork tools

Fieldwork tools are as ubiquitous as sites to investigate. Selection is determined by the purpose, location and data needs of the exercise. Selection also rests in the skill development you want for your students. Digital tools enable students to focus on understanding the meaning and purpose of data derived more than how to collect it.

Fieldwork tools

The skills students develop over time relate to the appropriate selection of a tool for the task at hand, effective use made of that tool, and the recording and transfer of the data it reveals. Some common fieldwork tools and their purposes are shown in Table 5.6.

Some tools are task specific (e.g. a rain gauge measures the amount of rainwater collected over time). Others are more multipurpose (e.g. an auger used for collecting soil samples can also be used for assessing vegetation matter in a sphagnum bog). Anything that enables collection of relevant data in the field can be classified as a fieldwork tool. Most tools are easily purchased, while some simple versions can be constructed by students in class prior to the fieldwork.

Table 5.6 Common fieldwork tools and their purposes

Tool	Purpose
Anemometer	Measure wind speed
Auger	Soil sampler (hand-held)
Barometer	Reads air pressure
Clinometer	Measures angle of elevation

Table 5.6 (*cont.*)

Tool	Purpose
Compass	Identify directions from and to features
Geological hammer	Enables taking rock samples (where permitted)
Gradometer	Measures slope gradients
Hygrometers	Measures the amount of water vapour in air, soil
Infiltrometer	Measures rate of water infiltration into soil
Light meter	Measures light luminance
Measuring tape	Measures distance and area
Pedometers	Counts steps (measure of relative distance)
pH meter	Soil and water testing
Quadrat frames	Assist counting of plants (or other feature) in an area
Rain gauge	Measures amount of rainfall
Ranging poles	Mark out key points for measuring, provides scale
Soil corer	Cylinder through which the core can be studied
Soil moisture readers	A probe measuring soil moisture levels
Sound level meter	measures noise levels in different environments
Stopwatch	Multiple uses – such as timing water velocity, traffic flows
Stream flowmeter	Measures rate of stream flow
Tally counter	Records tally counts (hand-held)
Thermometers	Measure air, soil or water temperatures
Topographic map	Indicates site location, relative elevation, and features
Trundle wheel	Measures distances as rolled from A to B
Turbidity tubes	Measures water clarity (suspended solids)
Vegetation identification books/kits	Easy-to-use guides to plant identification while in the field
Weathervane	Indicates wind direction

Fieldwork technology

The increased availability of digital and spatial technologies to support geography fieldwork in the collection of fieldwork data has been extraordinary in recent years. Technologies enabling field digital data collection include:

- **GPS.** The ability to log and geocode 'waypoints' in the field, contributes to their map and spatial analysis in class.

- **Smartphone apps.** Often, these are digital versions of 'old-school' fieldwork tools. Data capture and transfer can be an issue here.

- **Drones.** Collect remote sensed data, predominantly visual, previously not available during fieldwork.

- **GIS.** Field data loaded into a GIS in real time (laptop) enable gaps in data to be chased up while still in the field.

- **Internet.** Availability of widespread internet connection can provide in-field 'support' to assist data collection process (Holloway et al. 2020; IEEC 2021).

Some technologies may seem attractive to include in fieldwork but fall short of collecting primary data. They are more introductory familiarisation activities or ways to process and present the fieldwork results. These technologies include virtual maps, augmented reality, Google maps, stories or tour builders (Latham & McCormack 2007). As new digital technologies evolve, it is important that teachers are aware of and distinguish between those that 'collect' data and those that 'present' results. The latter can reduce the veracity of the actual 'fieldwork'. In assessing the usefulness of digital tools, teachers must consider the type and usefulness of the data collected, and where and in what format the data are retained.

The availability of smartphone apps and some software can also lead to difficulties of:

- provision – not all apps or software are available on different platforms

- access – the platforms students have available to them, raising equity questions

- compatibility – especially when involving the transfer of data and information.

The proliferation of digital and geospatial apps in smartphones offers great potential for invigorating fieldwork pedagogy; however, several limitations surrounding this technology exist, including:

- potential difficulties with, and equity of access to, the equipment

- steep learning curves associated with some digital programs and apps

- varying levels of student technical experience and expertise

- lack of research into best practice and methodology to maximise student learning (Holloway et al. 2020, p. 3).

Today's students have grown up with technology and evidence is growing that they 'have acquired distinctive new ways of learning' (Holloway et al. 2020, p. 2). Students are not put off by new technology and tools. Their approach to these tools is intuitive and requires little 'training'. Teachers' roles revolve around the assessment of the tools for gathering, storing and transferring appropriate data that meet fieldwork objectives (Holloway et al. 2020).

Short-answer questions

1.   Examine the list of fieldwork tools in Table 5.6, then list another five tools that could be included.

2.   Which of these tools (and others) could be constructed by students and used in the field?

3.   List geographic smartphone apps that you know of and note the skills students could develop. How could the teacher maximise students' skill development in using these apps?

Conclusion

The aim of the chapter has been to identify the core skills inherent in fieldwork. Fieldwork epitomises the role of geographical thinking that 'involves question-making [using] multiple resources to seek answers, analytic and synthetic thinking, and creative problem solving' (Kuismaa & Nokelainen 2018, p. 4). It is through careful planning and design of fieldwork, with considered thought given to the skills drawn upon, that this goal can be achieved.

Fieldwork skills are not topic specific. They should be seen as the glue that binds the techniques, methods and tools that we use in geographic fieldwork. Having clarity about the skills we are developing, whatever the context, will lead to more effective fieldwork and greater skill development of our students.

Bringing it together

1.   How important is the development of fieldwork skills to effective geography fieldwork?

2.   To what extent are the design and planning of fieldwork important in developing students' skills?

3. 'One could argue that "observation" is the central skill to all fieldwork.' Briefly outline the arguments for or against this statement.

4. What challenges are posed by the rapid development and availability of fieldwork technology for the development of students' fieldwork skills?

5. In what way(s) does the development of students' fieldwork skills and associated geographical thinking contribute to their geographical learning?

References

Anderson, J. & Erskine, K. (2014) Tropophilia: A study of people, place, and lifestyle travel, *Mobilities*, 9, 130–45.

Anderson, L.W., Krathwohl, D.R., Airasian, P.W. . . . Pintrich, P.R. (2001) *A taxonomy for learning, teaching, and assessing: A revision of Bloom's taxonomy of educational objectives*, New York: Longman.

Caton, D. (2006) Real world learning through geographical fieldwork, in D. Balderstone, ed., *Secondary geography handbook*, Sheffield: Geographical Association, pp. 60–73.

Esteves, M.H., Hortas, M.J. & Mendes, L. (2018) Fieldwork in geography education: An experience in initial teacher training program, *Didáctica Geográfica*, 19, 77–101.

Field, S. (2009) Doing fieldwork with Year 7 classes, *Geography Bulletin*, 41(1), 3–6.

Foskett, N. (2000) Fieldwork and the development of thinking skills, *Teaching Geography*, 25(3), 126–9.

Holloway, P., Kenna, T., Linehan, D., O'Connor, R., Bradley, H., O'Mahony, B. & Pinkham, R. (2020) Active learning using a smartphone app: analysing land use patterns in Cork City, Ireland, *Journal of Geography in Higher Education*, 45(1), doi:10.1080/03098265.2020.1802703

Holton, M. (2017) 'It was amazing to see our projects come to life!': Developing affective learning during geography fieldwork through tropophilia, *Journal of Geography in Higher Education*, 41(2), 198–212.

Illawarra Environmental Education Centre [IEEE] (2021) *IEEC geographical toolkit*, Wollongong: NSW Department of Education, retrieved from https://illawarra-e.schools.nsw.gov.au/resources-.html

Kuismaa, M. & Nokelainen, P. (2018) Effects of progressive inquiry on cognitive and affective learning outcomes in adolescents' geography education, *Frontline Learning Research*, 6(2), 1–19.

Latham, A. & McCormack, D.P. (2007) Developing 'real-world' methods in urban geography fieldwork, *Planet*, 18(1), 25–7.

Nowicki, M. (1999) Developing key skills through geography fieldwork, *Teaching Geography*, 24(3), 116–21.

Peasland, E.L., Henri, D.C., Morrell, L.J. & Scott, G.W. (2019) The influence of fieldwork design on student perceptions of skills development during field courses, *International Journal of Science Education*, 41(17), 2369–88.

Rydant, A.L., Shipleee, B.A., Smith, J.P. & Middlekauff, B.D. (2010) Applying sequential fieldwork skills across two international field courses, *Journal of Geography*, 109, 221–32.

Sharma, G. (2017) Pros and cons of different sampling techniques, *International Journal of Applied Research*, 3(7), 749–52.

Tuan, Y.F. (1974) *Topophilia: A study of environmental perceptions, attitudes, and values*, Chichester: Columbia University Press.

Wall, G. & Speake, J. (2012) European geography higher education fieldwork and skills agenda, *Journal of Geography in Higher Education*, 36(3), 421–35.

Whalen, K. & Paez, P. (2021) Student perceptions of reflection and the acquisition of higher order thinking skills in a university sustainability course, *Journal of Geography in Higher Education*, 45(1), 108–27.

Wheeler, A., Young, C., Oliver, K. & Smith, J. (2011) Study skills enhancement through geography and environmental fieldwork, *Planet*, 24(1), 14–20.

WJEC GCSE Geography (n.d.) *Fieldwork enquiry for Eduqas GCSE Geography A*, retrieved from www.hoddereducation.co.uk/media/Documents/Geography/9781471887406_MRN_WJEC-Geography-Fieldwork.pdf?ext=.pdf

Woolhouse, J. (2016) Case study of progressing geography fieldwork skills over Years 7–10, *Geographical Education*, 29, 40–5.

Using spatial technology

Mick Law

Learning objectives

By the end of this chapter, you will:

- recognise what geospatial technologies are and identify the different types of geospatial tools

- understand the benefits of using geospatial technologies in the classroom

- understand which geospatial tools are suitable for education and how best to integrate them into your teaching.

Introduction

With the digital revolution of the turn of the century, fast internet, widespread digital and mobile devices such as tablets, smart watches, mobile phones and the increasing functionality of the web-based tools we use have combined to bring some amazing geographical technologies into our homes and classrooms. Tools such as Google Earth, handheld GPS units and mobile smartphones have changed the way geography teachers bring the world to students. The smartphone, tablet, virtual reality (VR) and whatever technology is coming next will continue to make the subject even more relevant and useful to students in the future.

However, while knowledge *about* the geospatial industry and its tools is increasing, it is fair to say that the *implementation* of geospatial tools into our classrooms has been lagging. In the early part of this century, geospatial tools were clunky, difficult to navigate and usually the industry standard, which could be challenging for teachers who don't have time to learn new software. Consider how overwhelming it was the first time you ever opened a piece of industry-standard photo or video editing software and then add a full teaching load onto that anxiety!

Now these tools are far easier to use and to learn to use. The geospatial industry is moving more and more resources to the cloud; these include software and data that increase people's ability to use the tools. At the same time, there is now a mandate to use these tools in the classroom. The Australian Curriculum: Geography and all state syllabi require geography students to use geospatial technologies from early primary school onwards, so interest in geospatial tools is at its highest ever, and these tools will only become more widespread in geography as these trends continue.

All this background information brings us to you and your geography teaching. Through this chapter, we will explore the following questions:

- Why should you use geospatial tools in your classroom?

- What geospatial tools are appropriate?

- What will you do with your students to help them effectively use these tools to enhance their geographical learning?

- How are you going to incorporate geospatial technologies into your teaching craft sustainably?

What are geospatial technologies?

It seems very appropriate to examine the origins of the term 'geospatial' to help understand what the word means.

> Geo – a combining form meaning 'the Earth'
> Spatial – of or relating to space; existing or occurring in space.

The field of geospatial sciences is involved with the space related to the Earth; specifically, geospatial technologies are tools that help us to measure and represent the Earth. We tend to consider three main areas when we look at geospatial technologies, all of which will be discussed below. The main technologies related to the geospatial field are:

- geographical information systems (GIS)

- global navigation satellite systems (GNSS)

- remote sensing.

Naming conventions

One aspect of the geospatial sciences that can make the field hard to access is the inconsistent use of terminology across different parts of the geospatial community.

In the Australian Curriculum: Geography, the term 'spatial technologies' is used. In the industry, the term 'geospatial' is preferred to help highlight the geographic aspect of the information. Throughout this chapter, both terms will be used interchangeably.

Global navigation satellite systems (GNSS)

The first technology that should be introduced to students and teachers is the one with which they are generally most familiar: global navigation satellite systems (GNSS) technology, which most people refer to as 'GPS' (global positioning system).

As indicated, the naming can be confusing here: the United States began to put up the GPS network in the 1970s. This technology was only able to be used by the United States and its allies' military forces until the US Government opened access for the public in the late 1990s. This exclusivity forced other nations to create their own networks, such as Russia's GLONASS, Europe's Galileo and China's BeiDou; these, along with GPS, are all types of GNSS. Colloquially, these networks are all usually referred to as 'GPS', even though GPS specifically relates to the US network. There was limited civilian use (e.g. civilian air flights) before the late 1990s.

Pause and think

Consider the impact GNSS/GPS has had on your life. List the different ways you use GNSS/GPS.

How does it work?

GNSS consists of a network of satellites that orbit the Earth. Each satellite is steadily (and constantly) releasing pulses of radio energy that are picked up by various ground receiving stations or other receivers such as handheld GPS devices or modern mobile phone devices. The radio information transmitted includes the precise location of the satellite when the data were transmitted (in relation to the centre of its orbit around the Earth) and the exact time the pulse was transmitted. The receiver can then calculate its distance from the satellite by factoring the time taken for the data to be transmitted from satellite to receiver. Once the receiver has connections with three or four satellites, the receiver's position can be 'triangulated' and recorded, creating more connections and a stronger signal. This will mean greater accuracy in the position supplied by the receiver. The signal strength can be affected by interference if the user is close to large buildings or under heavily forested canopy. Common smartphones and handheld GPS receivers typically have +/–5 metre accuracy, which means the position as measured by the GPS receiver could be anywhere within a 5 metre radius of the actual position being measured.

How is it used?

From its origins as a tool of the military, GNSS now has many widespread general and commercial uses. Most commonly, these tools are used to locate and navigate. A range of recreational uses exist, such as geocaching, navigating outdoors and certain location-based augmented reality games such as Pokémon Go or Ingress Prime, which require the user to interact with the real world in order to influence their 'game world'. Hobbyists and professional sportspeople use these tools as personal fitness devices to help record their running or cycling tracks and additional information such as speed, acceleration, location and altitude. This information can then be shared, compared, mapped and analysed.

Geographic information system (GIS)

What is it?

As a broad concept, GIS is a framework for the capture and analysis of geographic and spatial data. GIS software allows the user to collect, represent, manipulate, query and analyse geographic and spatial data.

How does it work?

A GIS is a software tool that starts with a database of information. This database can also be represented on a map as each row of the database relates to a different location – a point, line or shape. Each column in the database contains a different piece of data, also known as attribute data.

Users can create or edit data in the database and then view that information on their map. The user can overlay multiple layers to examine patterns within and between layers of data and they can filter the data to look for particular information.

The power of GIS comes from the fact that each layer of information is connected/related to the database (the 'information system') and the user can overlay multiple layers of information and explore, query, analyse and manipulate the underlying data of and between those layers to give them a uniquely spatial perspective on their problem.

How is it used?

GIS originated as a natural resource-management tool. It was used for a range of things, including water quality monitoring, soil quality analysis, species tracking, vegetation identification and management, and water management.

GIS is a visualisation tool. Unlike an atlas, where the user can only view static information, GIS users can choose where to view and at which scale, and they can even decide which colours and symbols they see. GIS also allows the user to add, remove or change layers of data that are visible on their map. GIS is widely used as a cartographic tool to create thematic and other representations of the Earth.

GIS is used to analyse data. It can give users powerful tools to analyse their data such as representation tools (e.g. buffering), spatial analysis tools (e.g. heat mapping), and modelling and raw data analysis tools. Analysis can also occur when users query the data behind the map. They can filter and sort the 'attribute data' – the data that sit behind the map – as well as query data *between* layers on the map.

However, other industries have appreciated the power of spatial data and have adopted GIS as a data-analysis tool. This includes a broad range of industries such as health, town planning, transport, finance, banking, disease control, agriculture, marketing and engineering.

CONNECTION

John Snow's famous map

In 1854, London experienced a cholera outbreak that ended up killing thousands of its citizens, 616 of whom were in the Soho area of the city. Cholera is usually spread by contaminated food and water, although at the time of the outbreak the germ theory of disease was yet to be proposed so there was no way to actually know how the disease was being spread as it moved through the population. The medical community believed that the disease was being spread through the air, so efforts to study and contain the outbreak focused on this fact.

Soho, which had recently experienced a large influx of people, was a bustling corner of the city full of animal yards, slaughterhouses and grease-boiling merchants. London's sewerage system was yet to reach Soho, so the waste from all these businesses would pool and gather under floorboards and in gutters before slowly seeping into the ground.

John Snow, a physician and epidemiologist, examined the issue in Soho and took the unusual step, at the time, of mapping the data to examine any potential patterns. As it happened, this visualisation was a masterstroke and showed very apparently where the centre of this outbreak was located: the Broad Street water pump. Mapping the data (Figure 6.1) also highlighted a number of interesting outliers, including a business right next to the pump that had zero deaths from the disease and one victim who lived well away from the pump at the centre of the outbreak. The nearby business had its own water supply, meaning workers weren't exposed to the infected water being pumped from the Broad Street pump. The victim who lived away from the outbreak had her water delivered from that pump as she preferred its taste.

In the end, the handle of the pump was simply removed and the outbreak ceased a few days later. Snow went on to undertake further studies on London's water supply, which led to sweeping changes to the way water was managed and delivered to the population as well as a better understanding of cholera transmission. His lasting legacy, though, is using data visualisation to address challenges in society.

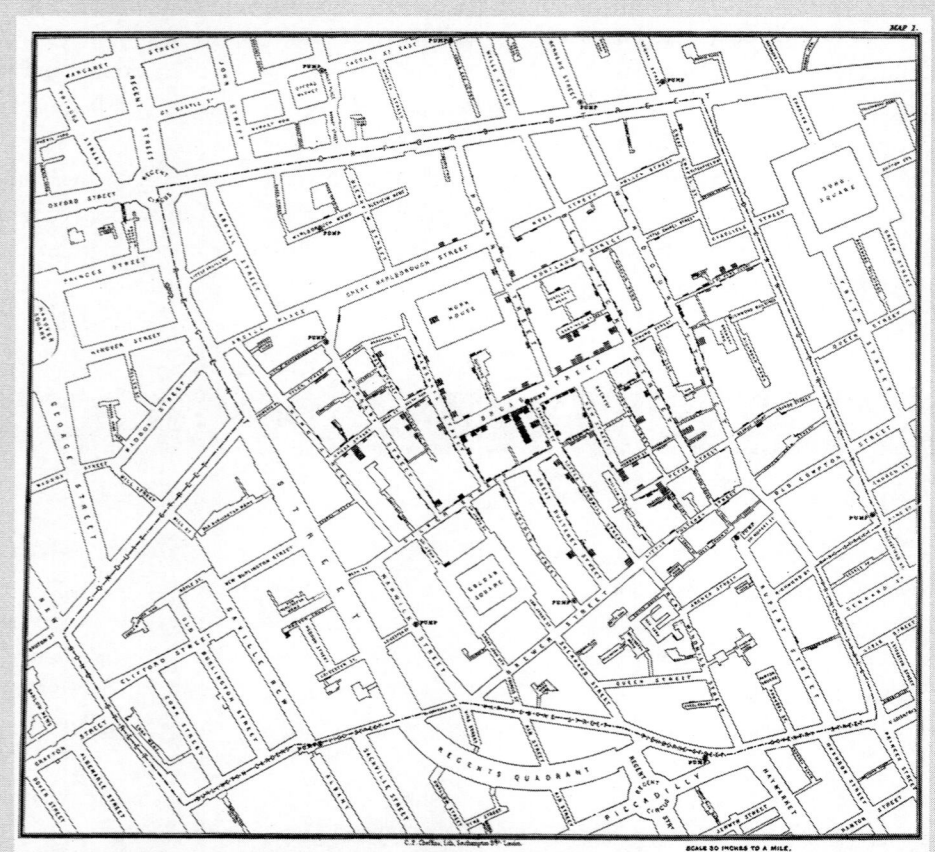

Figure 6.1 Snow's 1854 map of cholera

Questions

1. How are deaths represented in Snow's map?

2. What are the pros and cons of presenting cholera death data visually as opposed to using a table of information?

3. What other issues could be analysed using this technique?

4. How could you adapt Snow's map to make it even more effective?

Remote sensing

What is it?

The field of remote sensing constitutes data collected about the Earth while not being in physical contact with the Earth.

Remotely sensed data usually take the form of images taken by sensors mounted on drones, airplanes and satellites. Early twentieth-century technology included pigeons fitted with cameras that took images of the Earth from above, which would have been an incredible technological achievement at the time.

The outputs from remote sensing can be as simple as a satellite or aerial image of a place that uses the visible light spectrum as our eyes would see it, right through to imagery that visualises information from outside the visible light spectrum, such as ultraviolet or infrared light. These outputs are usually presented in the form of digital images made up of pixels, where each pixel has a colour value determined by the sensor or a combination of values from multiple sensors.

How does it work?

The sun emits energy, which travels to and bounces or reflects off the Earth. The energy that is reflected off the Earth reflects across a range of different wavelengths, and this energy can be measured by sensors. Different wavelengths of information can tell us different things about our world. Drones, aircraft and satellites can all mount different sensors as required for the project at hand.

Pause and think

1. List all the different ways in which we can remotely collect data about the Earth. How do you use this information in your day-to-day life?

2. Consider how you might teach with and about these tools in the classroom. What might be some challenges you might face in the classroom?

How is it used?

The most common type of remotely sensed data in our world today is imagery of a place taken in the visible band of the electromagnetic spectrum. This imagery can come from satellites, planes or drones. Weather information such as cloud cover, rainfall, mean sea level pressure or wind speed and direction is also remotely sensed. More specific uses of remotely sensed data include vegetation monitoring, urban growth monitoring, water quality monitoring, erosion mapping, weather monitoring, water level monitoring, light detection and ranging (Lidar) mapping, landslip monitoring and bushfire/thermal monitoring.

Short-answer questions

1. List the ways in which geospatial tools directly or indirectly influence your life.

2. How do you think you could use remotely sensed data in the classroom?

3. What data that you would collect in the field would be suitable for mapping using a GIS?

The benefits of using geospatial technologies in the classroom

What does the research say?

Geospatial technologies are instinctively recognisable to geography teachers as being beneficial in the classroom, even to those teachers without much hands-on experience with the tools. Tan and Chen (2015) note that geospatial technologies fall short of their potential to transform teaching and learning, specifically in the fields of social studies, geography, history and the sciences. At first glance, they appear to replicate traditional geographical tools, such as an atlas, but with greater accessibility, functionality and usability. But beyond a gut feeling, what do we know of the actual effectiveness of these tools?

There is a growing body of research supporting the benefits of geospatial and evidence that geospatial technologies have the power to be transformative tools for geography teachers in their classrooms (discussed in this section of the chapter). Most of the research indicates that geospatial tools:

- enable and enhance visualisation and promote spatial thinking
- help students understand geographical content
- enhance inquiry in both geography and science KLAs/subject areas
- increase motivation to learn.

Enhancing visualisation and promoting spatial thinking

The backbone of geospatial technologies is visualisation – having data in tables is one way to organise information, but being able to visualise geographic patterns is what sets GIS apart from other digital technologies. Kidman and Palmer (2006) determined that GIS promotes spatial thinking involving concepts of space, visualisation and [spatial] reasoning. Baker et al. (2015) and Demirci (2015) found that geospatial technologies help students to visualise spatial relationships and geographic patterns in their data, which allows them to better understand and analyse the data they represent.

García de la Vega (2019) highlights the link between geospatial technologies, visualisation and analysis of geographic patterns in data, while Baker et al. (2015, p. 121) note that, 'GST facilitates data collection, visualization of spatial relationships, analysis, and filtering or querying of geospatial data, all activities that can be of use in making sense of spatial data and patterns.'

Helping to understand geographical content

Students' understanding of geographical content is enhanced greatly by geospatial technologies. Goldstein and Alibrandi (2013) found that using geospatial technologies in the classroom had a significant effect on high school student achievement in science and social studies. Uttal et al. (2013) conducted a meta-analysis and determined that geospatial tools help to develop accurate understandings about complex Earth and environmental science concepts in secondary learners and a spatially enriched curriculum helps to increase performance and participation. Hammond et. al. (2018) found that geospatial technologies help to deepen students' understanding of important discipline-based content and complex Earth and environmental science concepts in secondary learners.

Enhancing inquiry

Geospatial tools are useful additions to a geographic inquiry for their data collection, representation and analysis abilities. By their nature, geospatial tools allow students to explore, analyse and question geographic data, all of which are useful attributes for a good geographic inquiry.

Geospatial technologies can play an integral part in a geographic inquiry as they are student centred (Kerski 2003) and allow students to actively explore their own understanding by learning through their own experiences (Demirci 2015). They can enhance geographic inquiry by 'allowing learners to formulate geographic questions or hypotheses, access and obtain geographic data from multiple sources, present geographic data and information in forms of maps, images, tables, and charts, explore the data through carefully constructed queries, and analyze the data to answer the questions or draw conclusions' (Hammond 2018, p. 14).

Increasing motivation

Anyone who has been in a typical classroom will have experienced and will understand the effect that using technology can have on student motivation to learn and to engage in learning. Hammond (2018), Kerski (2003), Demirci (2015) and Bednarz and Kemp (2011), among others, have found that using geospatial technologies in typical secondary school classrooms can increase student motivation considerably.

As this body of research continues to grow, it is likely that more findings will come to light that unveil the features of geospatial tools and pedagogy with regard to using the tools that provide benefit to students in the classroom. Teachers will benefit from more precise research on the pedagogies associated with teaching students how to use geospatial tools and how they are used in the real world. In putting out a call for a refined, coordinated research agenda in the field of geospatial technologies, Baker et al. (2015, p. 118) note that 'knowledge around geospatial technologies and learning remains sparse, inconsistent, and overly anecdotal'. Ground has been made up since then, but this highlights the relative immaturity of and opportunity in the field.

Short-answer questions

1. What are the obstacles to research in the field of geospatial education?

2. What benefits would using geospatial tools bring to your classroom?

Strategies to effectively bring geospatial technologies into your classroom

Learning about different tools, learning how to use them and learning how to use them *in the classroom* are very different things that bring their own challenges for a teacher. Geospatial tools are no different here to any other technical tools that you might bring into your teaching, and appropriate strategies must be employed to make sure you are using the tools to their fullest potential and not just as time-fillers or time-wasters.

Learning frameworks

There are a number of useful frameworks to consider when you are ready to properly incorporate geospatial technologies in the classroom. This chapter will focus on two different frameworks that can be used independently or together to help you enhance your teaching practice with these tools: TPACK and SAMR.

TPACK

The first is the technological, pedagogical and content knowledge framework that is also known as TPACK. The TPACK framework (Figure 6.2) emphasises the combination of content knowledge, pedagogical knowledge and technological knowledge in determining the success of pedagogical practice with technology in the classroom.

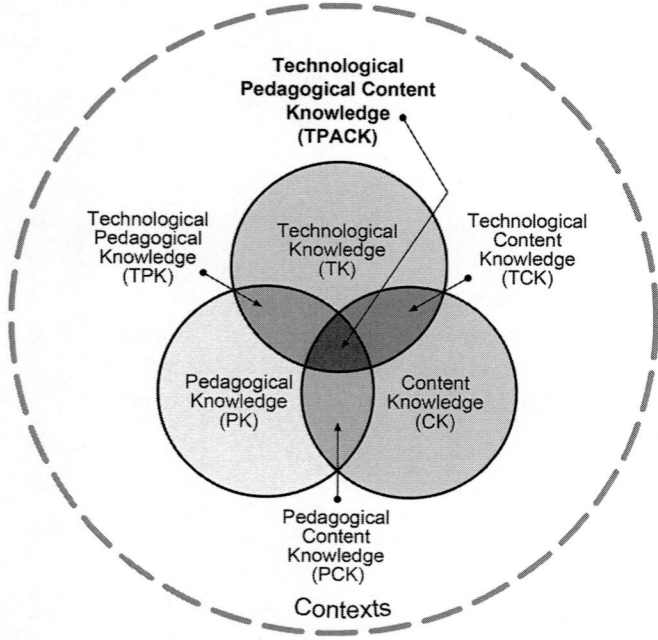

Figure 6.2 The TPACK framework

The TPACK model came about by adding technological knowledge to Shulman's (1986) PCK framework, which begins to describe how teachers bring content knowledge together with their pedagogical skills to teach students about and with technological tools to effectively teach with and about technology. In the context of geospatial technologies, technological knowledge encompasses knowledge about the use of geospatial tools – how to use them, what different buttons do, what different tools do, where different buttons are, outputs, data requirements and so on.

The pedagogical knowledge relates to knowledge and understanding of how to teach; the processes and practices of teaching and learning; and the practical skills used to help more effectively transfer knowledge to a person.

Content knowledge relates to teachers' knowledge about the particular geographical subject matter or issue being studied, such as earthquakes, land cover, volcanoes, water quality and so on. Shulman (1986) notes that this content knowledge includes 'knowledge of concepts, theories, ideas, organisational frameworks, knowledge of evidence and proof as well as established practices and approaches to developing that knowledge' (Koehler & Mishra 2009, p. 63).

Success will come when all three forms of knowledge are mastered and combined effectively in the classroom with the tools.

SAMR

The substitution, augmentation, modification, redefinition (SAMR) model (Figure 6.3) is another useful framework for examining how technology can be integrated into the classroom with increasing levels of sophistication, enhancement and transformative power for students.

The first two tiers of the model focus on how technology is a substitute for already existing knowledge/skills. The second two tiers focus on how technology can transform pedagogy and actually contribute new skills, knowledge, understanding, tasks and questions

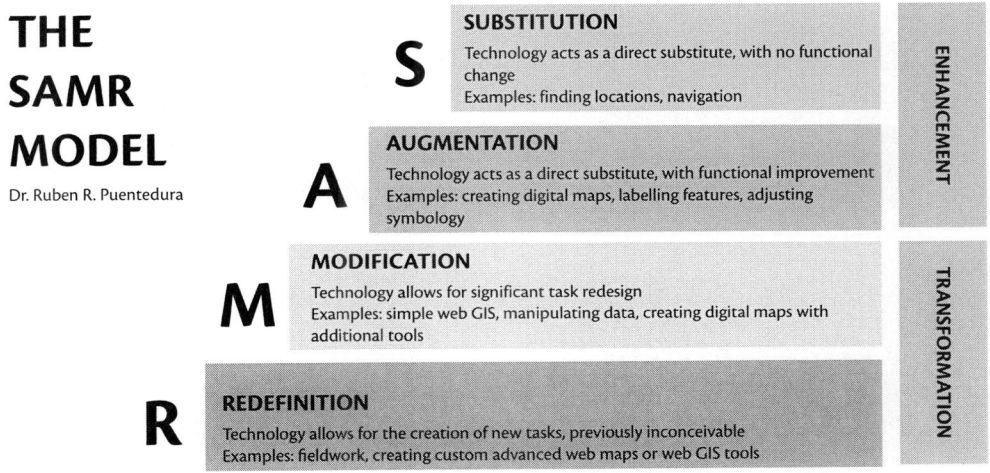

Figure 6.3 The SAMR model

in the classroom. It is interesting to note that there are opportunities across the SAMR model to incorporate geospatial tools with different levels of complexity and depth.

CONNECTION

Planning to use geospatial with TPACK and SAMR

Aamer's class is currently covering natural hazards, and he would like to get his students using some geospatial tools to emphasise their classroom learning about the distribution of bushfires, the geographical features that favour the bushfire hazard and their impact on Australian places.

He has gathered together a bunch of resources that he thinks might be useful, but he wants to consider them in line with TPACK and SAMR to make sure they will be effective tools for his students. The one he will focus on first is the DEA Hotspot map (https://hotspots.dea.ga.gov.au), created by Geoscience Australia, a federal government agency.

First, he jots down the different types of knowledge he will need to consider as he uses the tool in the classroom. He wants to see whether he has any gaps and where they are. He notes down some points, shown in Table 6.1.

Table 6.1 TPACK notes for the DEA Hotspot tool

	DEA Hotspot tool
Technical knowledge	• How to access the DEA Hotspot tool • How to bring up data • How to add layers to the map • How to view land cover data • How to view population data • How to view topographic data • How to view land use data
Content knowledge	• How bushfires start • Weather and climatic conditions that lead to bushfires • Topography that facilitates bushfires • How bushfires impact the environment • How bushfires impact people • How bushfires impact the economy • How we manage bushfire hazards

Table 6.1 (*cont.*)

	DEA Hotspot tool
Pedagogical knowledge	• How to use the student tablets/iPads • How to manage the class while they use the tablets/iPads • How to demonstrate to the students how to use the tools • How to keep students engaged and on-task while using the digital tools

Now he can see what he needs to know, he can also see that his technological knowledge is lacking, as there are things he wants his students to do that he just can't quite figure out in the software itself. He will therefore have to dedicate some time to playing with the software to figure that out.

Now he will use the SAMR theory to analyse the DEA Hotspot tool to try to consider all the ways he could use the tool in the classroom. It is really an opportunity for him to check that he is using the tool to its full potential and not just as a filler or time-waster. He notes down his ideas, shown in Table 6.2.

Table 6.2 SAMR notes for the DEA Hotspot tool

	DEA Hotspot tool
Substitution	• View where bushfires are occurring • View population information to see crossover
Augmentation	• View population and bushfire data on the same map using separate layers • Overlay land use, current weather or other data as required • Measure the distance from any current bushfire hotspots to populated places in Australia
Modification	• Measure the distance from any current bushfire hotspots to populated places in Australia • View bushfire events from a specific range of dates in a specific part of the country (e.g. Black Saturday in Victoria in 2009)
Redefinition	• May need to think here . . .

Questions

1. Pick a digital classroom tool with which you are already familiar and consider it using TPACK. What insights do you get from doing this?

2. In this example, the tool didn't really offer any opportunity to 'redefine' the task according to the SAMR model. Can you think of any examples of a digital tool that allow the user to redefine what they are doing? Explain how.

3. The example provided is an analysis of one tool using the two models. How could you implement the TPACK and SAMR models at a broader level – say, across your entire work program? What would be the value in doing that?

Map skills and start simple

It is worthwhile spending time mapping the geospatial skills that you want to teach your students across your work program to ensure you provide your students with appropriate exposure at the right time in their school life.

Allocate tools first where they are needed in the curriculum; this will flow from your geospatial skills mapping. Consider the geospatial skills required to be successful in senior geography and work backwards to ensure that your junior secondary students are exposed to those skills early on in their school lives. Make the use of geospatial tools as problem-solvers normal for students so they turn to these tools when analysis and decisions need to be undertaken.

Senior students need to collect their own data, manipulate the data and represent the data so they are able to be used for a specific purpose. Senior students are required to conduct more rigorous analysis of data and to manipulate their data meaningfully. Senior geography students also need to identify which data will be most useful for solving their problem as well as actually collecting their own data in the field. Bear all of these in mind as you map and apply skills to different areas of your work program.

What resources are there to get started?

When you are ready to incorporate geospatial tools into your classroom practice, you can take advantage of a range of freely available tools. Figure 6.4 shows the main categories of geospatial tools available to teachers. The largest and most accessible group is the online geospatial tools. Interactive geospatial tools are more sophisticated and require a good understanding of the fundamentals of geospatial data, and industry-standard tools are specialist and can be quite technical but are still useful in the classroom.

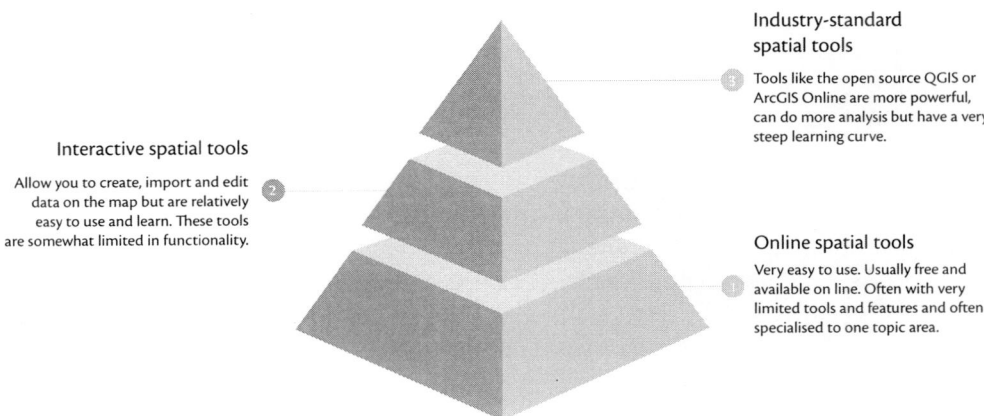

Interactive spatial tools

Allow you to create, import and edit data on the map but are relatively easy to use and learn. These tools are somewhat limited in functionality.

Industry-standard spatial tools

Tools like the open source QGIS or ArcGIS Online are more powerful, can do more analysis but have a very steep learning curve.

Online spatial tools

Very easy to use. Usually free and available on line. Often with very limited tools and features and often specialised to one topic area.

Figure 6.4 Types of geospatial tools

Once you have mapped geospatial skills across your work program, you will have an idea of what tools you need to use in each of your units of work and can begin to incorporate geospatial tools where they are appropriate.

Pause and think

1. What online geospatial tools have you used to date?

2. How could you use those tools in the classroom?

Online geospatial tools

Online geospatial tools are the best place to begin with geospatial skills in the classroom, as they are accessible, convenient, simple, focused on one theme and easily adaptable for classroom use.

Accessible

Obviously, online geospatial tools are available on the internet. This means they are available to anyone with an internet connection and a web browser. As web design has become more standardised across the internet, these tools have become simpler and easier to use. Rather than trying to pack too much data into a web-based geospatial tool, these tools tend to have limited data sets, usually based around a single theme such as a natural hazard, type of disease or global biomes.

Simple

As the internet has become increasingly integral to our society, its use and function have been standardised according to needs, security, accessibility and utility. Similar standardisation has occurred in online geospatial tools, making them more simple, accessible, functional, useful and user friendly.

Functional

Online geospatial tools are usually built around a narrow theme such as earthquakes, weather or migration, and come with a limited set of tools to help the user visualise and analyse the information on the site. With all data and analysis tools in one location, online geospatial tools are functional and almost optimised for classroom use.

Apps

Various apps offer teachers of all skill levels some usefulness in the classroom and in the field. Most 'geographical' apps involve some sort of movement. Here are a few worth-while apps you can download:

- navigation apps such as Google Maps; as well as navigating, you can collect and share GPS locations using Google Maps

- GPS logging or tracking apps (as opposed to GPS navigation apps, which will bring up turn-by-turn car navigation apps)

- names such as Geocaching, Pokémon GO or Ingress that require the user to navigate around the real world to play in the virtual game world.

More complex geospatial tools

A range of more substantial platforms are useful in the classroom. These tools tend to be more complex, but also more powerful. These tools often allow the user to include their own data in visualisations and to conduct more complex analysis of the data. The tools indicated below are all freely available online from within a web browser.

Google Earth

Google Earth is probably the world's best known geospatial tool, and has recognition almost everywhere. Google Earth revolutionised geospatial technology, not just in education, but by making geospatial information instantly accessible to almost everyone. Google Earth is available as a freely downloadable desktop application or via a simpler web-based application with less features. Google Earth is packed with data, very easy to use and full of features that are very useful for classroom teachers. It is highly recom-mended for use in the classroom.

Google Maps

Google Maps is probably the world's second best known geospatial tool and one that was made here in Australia. Google Maps is available via a web browser or as a widely popular smartphone app. The tool lets the user navigate and also see a lot of data about the world, including topography, traffic data, public transport times, satellite imagery, and bike paths and tracks. Users can sign into Google My Maps to create their own custom maps using their own spatial data.

NASA WorldWind

NASA's WorldWind has been around for a long time, first launching to the public in 2003 (before Google Earth). The web-based tool allows the user to view some data, but its power comes when users include their own data in the visualisations. Everything about WorldWind is open source, so your students who are proficient in programming could even customise the tool to include whatever data or interface they like.

Intermediate geospatial tools

There are some useful tools that might be described as intermediate geospatial tools, which students can take advantage of to extend their subject area work. Rather than simply visualising information around one particular theme, intermediate geospatial tools allow the user to input, import, manipulate, edit, represent and analyse their own data via the tool.

Industry-standard tools

Whereas the main commercial geospatial tools were expensive, they are now becoming more affordable. While they were for high-end desktop machines only, now they are accessible on many devices almost anywhere. While they were complex pieces of specialist software, they are now cleverly designed apps with simple interfaces (actually, some are still complex!) that are available online. Many geospatial software providers have developed online platforms that can handle large amounts of data, powerful analysis and a useful set of tools.

QGIS

Quantum GIS (QGIS) is worth noting as an open-source geospatial tool that is free to download and is compatible with many data types. QGIS has an open source licence, which means the core software can be shared and edited, and useful plugins can be made easily (for a programmer) and at no cost, apart from time. A lot of these plugins are freely available and let more specialised users do a wide range of specific and useful actions. QGIS comes with only basic data sets, but it allows the user to import data from a range of sources, including the local computer or the internet via WMS servers.

Data

Data is usually the hardest element to get for teachers who want to extend their work from simple geospatial tools to more complex tools discussed in this section. Often the data are hard to find and, once found, they might not be in the right format. They might be too old or incomplete for your study area so data can be challenging to source.

Recent moves to make government data more accessible to the public through Creative Commons licensing and robust metadata standards have seen more data, and data of better quality, being released to the public locally, in all states and federally. This is great for teachers who want to use data covering places relevant to their students – places local to them that they know and places around the country that they should know. It is also reassuring to know that the quality of the data is getting better and better.

Check with your local council, state government or the federal government's online data repositories for spatial data. You should be able to filter by data type, and there you can search through a range of spatial data types and services such as .kml, .shp, .gdb, spreadsheet or Web Mapping Service (WMS) to find what you need.

Formatting spatial data

Spatial data are uniquely formatted. At their core, spatial data contain locational information, which could be a latitude and longitude, country name, state or province, address, point of interest or some other kind of grid reference. Each row contains data for a different location.

In a typical spreadsheet (see Figure 6.5), each column or field contains a unique piece of information or attribute. A unique identifier plus the location information are the minimum requirements. More information can be added as required – this could include statistics, dates, people, places, names, survey results and other data collected in the field.

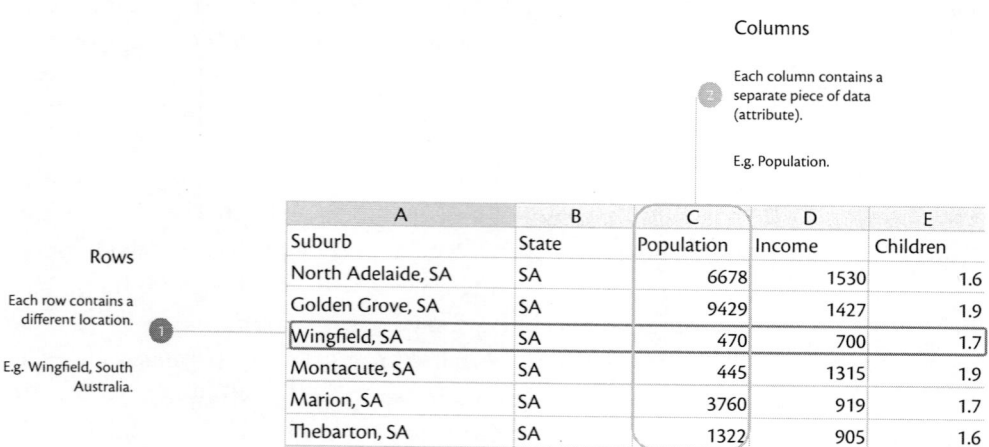

Columns

Each column contains a separate piece of data (attribute).

E.g. Population.

Rows

Each row contains a different location.

E.g. Wingfield, South Australia.

	A	B	C	D	E
	Suburb	State	Population	Income	Children
	North Adelaide, SA	SA	6678	1530	1.6
	Golden Grove, SA	SA	9429	1427	1.9
	Wingfield, SA	SA	470	700	1.7
	Montacute, SA	SA	445	1315	1.9
	Marion, SA	SA	3760	919	1.7
	Thebarton, SA	SA	1322	905	1.6

Figure 6.5 An example of a typical spatial data organisation table

Once you have collected your data, you will need to analyse them. To analyse your data, you will most likely need to represent them. The most common methods of data representation include charts, graphs, maps, tables, infographics and figures (see Chapters 3 and 4).

After your analysis, you will want to use the data to make some sort of difference in the world – this should bring you back to the original problem you identified before collecting your data.

Short-answer questions

1. What are the benefits of using the TPACK and SAMR models to audit your use of geospatial tools?

2. In what part of a geographical inquiry will geospatial tools be most useful?

3. What are the two components of spatial data?

Conclusion

Geospatial technologies are becoming more widespread and more useful to classroom teachers. After reading this chapter, you should have a good idea of what these tools are and what benefits they can bring to your teaching.

Geospatial technologies have the potential to transform your geography education if they are implemented appropriately. Consider your objectives in the classroom and how well the tool you want to use will complement what you wish to achieve. Apply the TPACK and SAMR models to give your implementation of technology real purpose.

Finally, play and make mistakes! Just playing with the tools means you have some experience that your students won't, and when they present you with a problem you may even have experienced it yourself.

Bringing it together

1. Describe the features of the three most prominent geospatial technologies.

2. What are the TPACK and SAMR models and how can they be applied to geospatial tools in the classroom?

3. List all the freely available tools that you could use to teach with and about geospatial technologies in your classroom. Categorise them according to their geospatial skills.

4. List as many careers as you can think of that use geospatial tools in some way.

5. Develop the outline of a geographical inquiry that uses geospatial technologies to collect, represent, analyse and communicate information. Consider a topic that you will be teaching and create inquiry questions, outline your methodology and your analysis, and decide how you will communicate your findings.

6. Choose a unit of work (note the topic and year level) that you enjoyed teaching while undertaking your practical teaching experience. Write or adapt a lesson for that topic that incorporates geospatial technologies in some way. Consider how you are using geospatial tools in relation to the TPACK and SAMR models covered during the chapter. If necessary, revise your inclusion of the geospatial tools to make them more effective.

References

Baker, T.R., Battersby, S., Bednarz, S.W., Bodzin, A.M., Kolvoord, B., Moore, S., Sinton, D. & Uttal, D. (2015) A research agenda for geospatial technologies and learning, *Journal of Geography*, 114(3), 118–30.

Bednarz, S.W. & Kemp, K. (2011) Understanding and nurturing spatial literacy, *Procedia – Social and Behavioral Sciences*, 21, 18–23.

Demirci, A. (2015) The effectiveness of geospatial practices in education, in O. Muñiz Solari, A. Demirci & J. Schee, eds, *Geospatial technologies and geography education in a changing world: Advances in geographical and environmental sciences*, Tokyo: Springer, pp. 141–53.

García de la Vega, A. (2019) Spatial thinking ability acquisition through geospatial technologies for lifelong learning, in R. de Miguel González, K. Donert & K. Koutsopoulos, eds, *Geospatial technologies in geography education: Key challenges in geography*, Cham: Springer, pp. 21–40.

Goldstein, D. & Alibrandi, M. (2013) Integrating GIS in the middle school curriculum: impacts on diverse students' standardized test scores, *Journal of Geography*, 112(2), 68–74.

Hammond, T.C. et al. (2018) 'You know you can do this, right?': Developing geospatial technological pedagogical content knowledge and enhancing teachers' cartographic practices with socio-environmental science investigations, *Cartography and Geographic Information Science*, 45(4), 305–18.

Kerski, J.J. (2003) The implementation and effectiveness of geographic information systems technology and methods in secondary education, *Journal of Geography*, 102(3), 128–37.

Kidman, G. & Palmer, G. (2006) GIS: The technology is there but the teaching is yet to catch up, *International Research in Geographical and Environmental Education*, 15(3), 289–96.

Koehler, M. & Mishra, P. (2009) What is technological pedagogical content knowledge (TPACK)?, *Contemporary Issues in Technology and Teacher Education*, 9(1), 60–70.

Shulman, L.S. (1986) Those who understand: Knowledge growth in teaching, *Educational Researcher*, 15(2), 4–14.

Tan, G.C.I. & Chen, Q.F.J. (2015) An assessment of the use of GIS in teaching, in O. Muñiz Solari, A. Demirci & J. Schee, eds, *Geospatial technologies and geography education in a changing world: Advances in geographical and environmental sciences*, Tokyo: Springer, pp. 155–67.

Uttal, D.H., Meadow, N.G., Tipton, E., Hand, L.L., Alden, A.R., Warren, C. & Newcombe, N.S. (2013) The malleability of spatial skills: A meta-analysis of training studies, *Psychological Bulletin*, 139(2), 352–402.

PART 3
TEACHING GEOGRAPHY

The inquiry process in geography

John Butler and Susan Caldis

Learning objectives

By the end of this chapter, you should be able to:

- understand the importance of inquiry in the Australian Curriculum: Geography
- understand why inquiry is a central process in learning
- identify a range of techniques that support and develop the process of inquiry in the teaching and learning of geography

Introduction

This chapter deals with inquiry, a central tenet of geographical learning as set out in the Australian Curriculum. It examines the reasons for using inquiry in the classroom, in terms of both pedagogical theory and practical classroom teaching. It then suggests a number of ways in which inquiry can be put into practice.

Inquiry as part of the Australian Curriculum

The Australian Curriculum: Geography gives the following guidance:

> Geographical inquiry is a process by which students learn about and deepen their holistic understanding of their world. It involves individual or group investigations that start with geographical questions and proceed through the collection, evaluation, analysis and interpretation of information to the development of conclusions and proposals for actions. Inquiries may vary in scale and geographical context. (ACARA 2020)

With this statement, the Australian Curriculum: Geography is placing inquiry in a central position in terms of the delivery of the curriculum to students. It is followed up by the linking of inquiry with the skills to be mastered.

Each year level includes 'key inquiry questions that provide a framework for developing students' geographical knowledge and understanding, and inquiry and skills' (ACARA 2016). The content and skills for each topic suggest ways in which the learning can be structured as part of an inquiry.

Each unit of the Australian Curriculum: Geography contains key inquiry questions that suggest the broad approach that should be taken. Within the Inquiry and Skills section of each unit, the following sequence of inquiry is spelled out:

- **Observing, questioning and planning:** Identifying an issue or problem and developing geographical questions to investigate the issue or find an answer to the problem.

- **Collecting, recording, evaluating and representing:** Collecting information from primary and/or secondary sources, recording the information, evaluating it for reliability and bias, and representing it in a variety of forms.

- **Interpreting analysing and concluding:** Making sense of information gathered by identifying order, diversity, patterns, distributions, trends, anomalies, gener-alisations and cause-and-effect relationships, using quantitative and qualitative methods appropriate to the type of inquiry and developing conclusions. It also involves interpreting the results of this analysis and developing conclusions.

- **Communicating:** Communicating the results of investigations using combinations of methods (written, oral, audio, physical, graphical, visual and mapping) appropri-ate to the subject matter, purpose and audience.

- **Reflecting and responding:** Evaluating findings of an investigation to reflect on what has been learnt and the process and effectiveness of the inquiry; proposing actions that consider environmental, economic and social factors; and reflecting on implications of proposed or realised actions (ACARA 2020).

Teachers need to be aware that each stage in the process of inquiry cannot possibly be covered completely in every task or activity. There would not be enough time to do so, and students would find the repetition of techniques cumbersome. Some inquiry activities are well-suited to using each stage in the process; however, other inquiry activities will need to select certain stages for emphasis. In some class activities, different groups of students may undertake different stages of the inquiry process, working with the same inquiry question and context or case study, then put all the work together late in the process. Figure 7.1 shows the flexibility of inquiry.

Inquiry is linked intimately with knowledge and understanding. In the best teaching and learning of geography, students are using the stages of inquiry to extend their

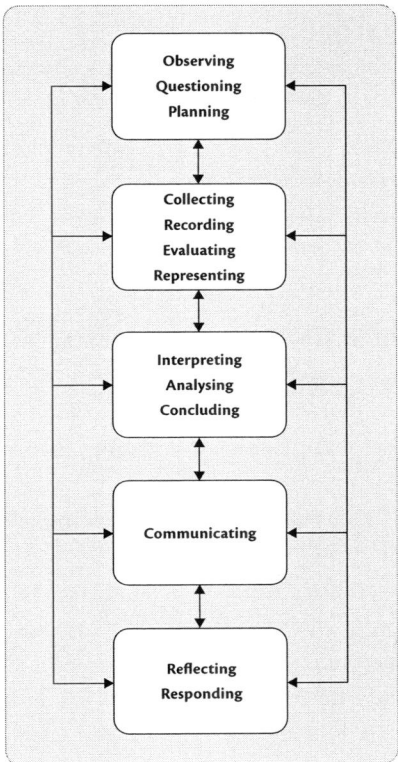

Figure 7.1 Possible flexible routes of the inquiry process

knowledge, deepen their understanding and develop skills that can enable them to think geographically. The Australian Curriculum: Geography states that 'Geography teaches students to respond to questions in a geographically distinctive way; plan inquiries; collect, evaluate, analyse and interpret information; and suggest responses to what they have learnt' (ACARA 2020).

Short-answer questions

1. How might Figure 7.1 be enacted in a geography lesson, or series of geography lessons? Would students have to start at Observing, Questioning, Planning? Justify your response.

2. Choose a unit of work from the Australian Curriculum: Geography and develop an alternative set of three inquiry questions to frame learning. How are the alternative questions similar to or different from those posed in the curriculum

document? Why are such questions important to you as an aspiring geography teacher responsible for teaching the given unit of work?

3. What part of Figure 7.1 do you believe to be most challenging to enact and why?

4. What inquiries might be particularly suitable for students in early secondary school, and for those in the final years of school?

Why is an inquiry approach used in teaching and learning geography?

There are a number of good reasons why inquiry is central to learning in geography. Margaret Roberts suggests four important reasons for using inquiry in her books *Learning Through Enquiry* (2003) and *Geography Through Enquiry* (2013).

First, geography is not a group of facts there to be learned. Like all subjects, it is a construction that has grown and changed over time. Geography has its roots in exploration and expanding knowledge of the world, and technology has contributed greatly to the change occurring in the subject and discipline, particularly in the way data and information are gathered and interpreted (see Figures 7.2 and 7.3). Geographical understanding has been constructed by its practitioners through the questions that have interested them and the techniques they have found to be most useful in answering these questions. Responses to such questions have changed from being descriptions about far-away places to focusing on analysing known places. The inquiry will draw on lived experience and the gathering of data and information from quantitative and qualitative methods with a focus on social justice and inequality. A focus on the experience of and immersion in place, space and environment, together with change over time and human interactions, means that geographical inquiries are often interpreted through a future-focused, action-oriented lens. Geographic techniques have changed from hand-drawn mapping and recording to the use of computers handling big data and to sophisticated spatial data available to all on smartphones.

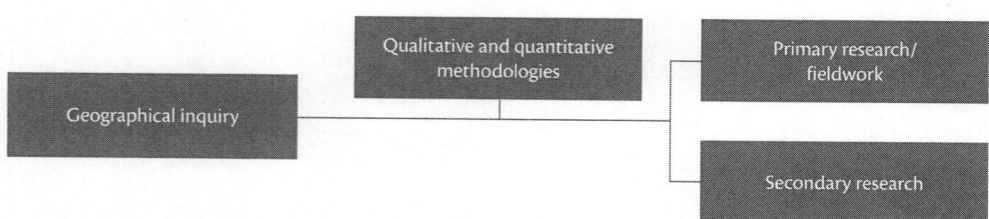

Figure 7.2 A geographical inquiry, part 1

Source: adapted from Caldis (2015).

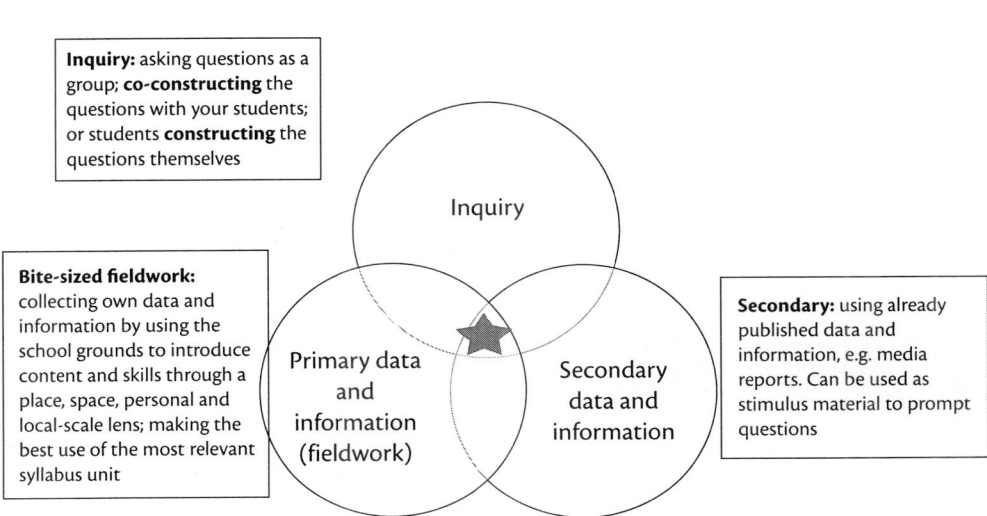

A geographical inquiry with bite-sized fieldwork and secondary research incorporated into one lesson or a sequence of lessons

Inquiry: asking questions as a group; **co-constructing** the questions with your students; or students **constructing** the questions themselves

Inquiry

Bite-sized fieldwork: collecting own data and information by using the school grounds to introduce content and skills through a place, space, personal and local-scale lens; making the best use of the most relevant syllabus unit

Primary data and information (fieldwork)

Secondary data and information

Secondary: using already published data and information, e.g. media reports. Can be used as stimulus material to prompt questions

Figure 7.3 A geographical inquiry, part 2

Source: based on ideas presented in Caldis (2019b).

A second set of factors relate to our knowledge of how children learn. Teachers understand that knowledge cannot be poured from a jug into an empty mind. They know that each child develops at a different rate, and makes sense of the world in their own way. A constructivist approach to teaching accepts that each individual has different expectations, experiences and attitudes. What they learn is individual to each person because they construct learning in relation to what they already know. We construct our understanding of the world differently from everyone else, and we keep changing it as we have new experiences and keep asking questions in response to the new experiences and our development in understanding.

A third reason why inquiry fits well with learning geography is the interdisciplinary nature of the subject. Geography looks at both the physical and human features of the world, and the ways in which they link together. Such linkages lead to development of 'wicked problems', and thus shape the questions that frame future-focused, action-oriented learning. Geographical understanding incorporates knowledge and techniques from geology, biology, oceanography, sociology, economics and ecology in a unique way. The interpretation of every issue studied by a geographer is influenced by an overlap of many of these disciplines, and is best understood through the posing of questions or problem-statements to form the basis of an inquiry. To use the example of geography and science, technology, engineering and mathematics (STEM), the discipline of geography straddles both the sciences, and the humanities and social sciences (HASS). In curriculum design and also in practice in schools, geography as a subject tends to be identified (and perhaps pigeonholed) in the HASS learning area. However, it is the inquiry method and fieldwork inherent to geography that become not only the distinctive features of the subject, but also its point of connection with science and the field of STEM. For the STEM field, geography adds depth to learning through its focus on place-based analysis, spatial reasoning and understanding the nature and effects of human–environment interactions more than human interactions. The authentic connection and point of intersection between geography and the STEM field occurs through inquiry, fieldwork and problem-solving with a sustainable-futures lens (Caldis 2019a; Caldis & Kleeman 2019).

A fourth justification for an inquiry approach is its suitability to develop a wide range of capabilities in students. The Australian Curriculum lists these desired capabilities as literacy, numeracy, ICT capability, critical and creative thinking, personal and social capability, ethical understanding and intercultural understanding (discussed in detail in Chapter 10). An inquiry approach to learning encompasses all of these.

CONNECTION

What do the education theorists say about inquiry?

Inquiry is associated with a constructivist approach to learning and the development of knowledge (Firth 2015; Stemhagen, Reich & Mutch 2013). Constructivism puts student beliefs, questions and lived experiences at the centre of learning, and in doing so simultaneously encourages active participation in learning from the students and a movement away from a hierarchically focused dissemination of information by teachers (Stemhagen et al. 2013). By its very nature, an inquiry approach towards teaching and learning requires curiosity and wonder to be awakened through asking and pondering questions (Spronken-Smith et al. 2008).

Inquiry fits well into a constructivist view of curriculum, but it also can be used in a curriculum that needs explicit instruction for some age groups, some topics and some skills. Bergsen et al. (2020) argue that students with prior expertise in an area can learn more efficiently in an open-ended inquiry, while those who are beginners in an area need a different kind of learning experience. For the teacher, this means using a mix of direct instruction and more indirect inquiry in the classroom to fulfil the needs of different students at different levels of skills mastery. Table 7.1 presents an example of what this might look like when planning a unit of work.

Table 7.1 An example to show explicit instruction is incorporated into an inquiry-based teaching, learning and assessment sequence of learning

Overarching question to frame the unit of work		
Lesson (L) 1: Introduction and co-construction of inquiry question	L2: Setting a plan for the conduct of a geographical inquiry (primary and secondary methodologies)	L3: Explicit instruction lesson (concepts, terminology, tools)

Table 7.1 (*cont.*)

Overarching question to frame the unit of work		
L4 & 5: Student reflection on and investigation (secondary) of the question to start hypothesising and responding to the question		L6: Students conduct bite-sized fieldwork (primary)
L7 & 8: Student reflection on gathered primary and secondary data and information in response to the question: What's this? What now? What next? What might? What can I do? What can be done for the future?		L9: Explicit instruction lesson (concepts, terminology, tools)
L10: Student investigation (secondary) to build response and student reflection on suitability of planned bite-sized fieldwork for next lesson	L11: Students conduct further bite-sized fieldwork (primary)	L12: Student reflection on and consolidation of data and information gathered in response to question
L13: Student reflection on and consolidation of data and information gathered in response to question	L14: Explicit instruction lesson (concepts, terminology, tools)	L15: Student investigation (secondary or primary as appropriate) to build and finalise response to question
L16–18: The overarching question is answered and communicated in a geo-literate way, either individually or collaboratively. Peer assessment could also occur.		

Short-answer questions

1. Which of the reasons for inquiry being central to geography learning do you find most compelling? Justify your response.

2. How would you explain to a student why geography uses inquiry is central to learning?

3. Read the article 'Fostering geographical inquiry and fieldwork', which can be downloaded from AGTA's online resources: www.agta.asn.au/Resources/ProfessionalStandards/index.php. How does the information related to this

GEOGstandard reinforce or challenge your thinking about the use of an inquiry approach in geography? Identify further questions that arise from your reflection on this chapter so far.

Inquiry and thinking

A very important – perhaps the most important – aim of schooling is to develop thinking skills in students. Encouraging inquiry helps students to deepen their understanding and enhance their intellectual engagement. Finding out answers and learning facts are not nearly as important as making sense of information and making connections. Teachers can create a program of inquiry that helps students to develop their thinking skills by using some of the techniques described below:

- Develop units of work that contain deep inquiry questions.
- Create a need to know by using stimulus material that appeals to the age group and interests of the class.
- Ensure that inquiry questions are related to students' prior knowledge, or suggested questions, local issues or current events. Any of these will help students to see relevance and purpose in their study.
- Encourage questions and inquiry about controversial issues.
- Show students that you are supportive of them as individuals, and particularly of their questions and ideas.
- Use scaffolding to show students possible structures of investigation, analysis and reporting.
- Encourage the sharing of ideas between students and groups.
- Use techniques that develop critical thinking and problem-solving.

Creating a need to know

If every teacher could stimulate inquiry in all students by pressing a small switch inside them that said to their brain, 'I need to know this', classrooms would operate like a magic kingdom! Perhaps we cannot do this, but fortunately we have a number of techniques that do encourage a 'need to know' in many students.

Possibly the most effective technique is for the teacher to be an inquirer, who gets personal pleasure from discovering new knowledge, and can share that pleasure with students. The teacher who often brings a new article from *National Geographic* or a film clip from YouTube into the classroom can encourage an interest in some students. This pleasure in discovery is illustrated by the way David Attenborough presents his fascination with the world around him. Classroom teachers could do a lot worse than adopting some of his approaches.

Sometimes creating a need to know can occur from looking out of a window. The window can be literal (such as a classroom window) or metaphorical (such as a photograph). The literal and metaphorical windows become points from which students pose questions in response to what they can observe – or not. Students can also draw on their lived experience and existing knowledge of a familiar area to develop questions. Upon looking out of a window, common questions to arise include: 'Where is this place?'; 'What direction am I facing?'; 'Why is X happening?'; 'What is X over there and why is it doing?' As a result, an inquiry approach becomes central to practice, and the opportunity to conduct a geographical investigation emerges. Looking out of a window will harness observation skills by using the senses (sight, sound, smell, feeling/emotion) and it will also prompt curiosity about what is happening now, whether it has always occurred this way and what might be next

Pause and think

Find your window! Identify your window into the subject of geography – what was it that sparked your curiosity and interest about the subject? Look out of a literal window – what do you observe and what questions do you have? How might you be able to use the idea of a 'window' with students in your geography classes?

Roberts (2013) suggests that speculation is a powerful way of stimulating a need to know. Students might be asked to speculate about:

- how a feature or characteristic of the Earth became what it is now

- how a question or issue might be investigated

- estimations of statistics, sizes or length of time.

For some students, a concrete object is an effective stimulus that promotes the need to know. Some suggestions for stimulus objects are:

- short film clips

- souvenirs or typical objects from particular places

- everyday objects that link the students to other places where they were made, grown or devised

- maps of different kinds – particularly mapping oddities, strange places, unusual patterns

- interactive websites such as and Worldmapper (https://worldmapper.org) and Gapminder (www.gapminder.org)

- vivid photographs – either on paper or digital.

Short-answer questions

1.   Suggest some examples of objects that have stimulated you to want to know more about them.

2.   As a geographer, you have seen many kinds of maps. Which maps excited you most and made you want to visit that place?

3.   What are some video clips or television programs that have stimulated your imagination about places and environments?

Using questions

Questions are a key ingredient of inquiry. At the start of each year level, the Australian Curriculum: Geography gives teachers three inquiry questions. These are broad questions that attempt to cover a large slice of the concepts and issues in the units for that year level. Teachers need to look carefully at the intention of these questions, and work towards producing questions with a finer focus as sub-sets of these, as well as stimulating students to devise their own questions.

Rawling (2007) suggests a framework of questions related to the different stages of the inquiry model shown above (Table 7.2).

Table 7.2 Question words that fit stages of inquiry

Questions	Inquiry skills involved
What?	Observation and perception
What and where?	Definition and description
How and why?	Analysis
	Explanation
What might happen?	Evaluation
What impact?	Prediction
What decision?	Decision making
What do I think? Why?	Personal evaluation Response
What will I do next?	

Source: based on the ideas presented in Rawling (2007).

Another area to consider with regard to questioning is the co-construction of an inquiry question. Co-construction means the teacher and the students collaborate to develop and agree on an inquiry question to investigate. If students are always presented with an inquiry question that is either straight from the curriculum document or from a textbook, or one the teacher has developed, there is no input from students. Students may not see a question that is presented to them as relevant to the unit under investigation – it is not part of their view from their 'window'. A literal window is recommended as a starting point for co-constructing a question and from here comes curiosity; however, if a literal window is not available, a metaphorical window needs to be provided. It is important to find your own literal and metaphorical window as a teacher and then use this, together with your knowledge about the school and student context, to find their window. It is best to find a classroom window – what is a point of interest for everyone? What do we all need to know and why? If student curiosity is evident in an inquiry question *at some point* in the teaching and learning process for a unit of work, then student voice is encouraged and an opportunity for enhanced levels of engagement occurs.

Co-constructing an inquiry question (1) gets students actively involved in their learning; (2) also helps students to see the point of what they are learning in geography; and (3) creates a shared understanding about the process of learning and content knowledge to be developed. The co-construction of an inquiry question allows a teacher to model the process of developing a researchable, geographical question and for students to participate in the process.

CONNECTION

A teacher's example

This example is drawn from one of the authors' experiences of teaching geography in a secondary school context. Although a response to the question is determined through fieldwork around the local area, it is the process of co-constructing a question that is the focus of learning in this instance. To co-construct a question, you need to allow students time to think and to look at something with which they are familiar. However, they need to consider it in terms of being a geographer – that is, through acknowledging the personal and local scale in the first instance, and through using the concepts of place, space, environment, interconnection, change and sustainability. The process for co-constructing a question is identified in Table 7.3.

Table 7.3 Steps in the process of co-constructing an inquiry question

Steps in the process of co-constructing an inquiry question
Before the lesson, as part of preparation:
1. Familiarise yourself with the curriculum and unit of work as appropriate (e.g. Stage 4, Place and Liveability); determine a suitable content description (e.g. Factors that influence the decisions people make about where to live and their perceptions about the liveability of places) to become the 'area of focus'

Table 7.3 (*cont.*)

Steps in the process of co-constructing an inquiry question

2. For the 'area of focus', there could be many potential questions (e.g. 'Why do people live in this suburb?'; 'What might encourage people to live in this area?'; 'Is <insert suburb name> considered to be a liveable suburb? Why or why not?'). As the teacher, you need to have an understanding of the nature of some possible questions and facilitate student responses towards this 'area of focus'.

3. Know what you would like students to do with the inquiry question once it is developed to help with preparation beyond the next lesson and to help refinement of a researchable geographical question (e.g. find some possible responses through localised, short fieldwork activities such as recording tallies, making observation notes, taking photographs and annotating the image, using apps or equipment to measure and record wind speed, temperature and noise levels, interview a known member of the community). Consequently, time, minimal equipment, concepts and the need to do fieldwork become the criteria against which the question is determined and agreed upon.

During the lesson:

4. Review the previous lesson, key terminology (e.g. meaning of liveability) and capture key learnings for easy viewing/access by students (e.g. write on the whiteboard); introduce the focus for this lesson (e.g. we know what liveability means, now we are going to explore the liveability of particular places such as the suburb in which we are currently located).

5. Invite three to five students to go and look out of the classroom window and articulate (1) what they expect to see; (2) what they can see; (3) what they cannot see; and (4) what questions they have. Capture their responses on the whiteboard using a different colour for each category of response. Students can be invited by name, or you can ask for three to five students to nominate themselves to go up to the window. Remember to pause and allow time for students to respond.

6. Drawing on the responses from students at the window, ask the rest of the class whether the responses are related to a place being liveable (yes, no, why/why not); depending on student responses, you may need to repeat steps 7, 8, 9. Remember to pause and allow time for students to respond. Invite students at the window to return to their seats and desk.

Table 7.3 (*cont.*)

Steps in the process of co-constructing an inquiry question
7. Drawing on the student responses from Steps 5 and 6 encourages the class to suggest possible questions (perhaps three to four questions), which could be investigated to find out more about the liveability of 'this' suburb or known place. Capture the responses on the whiteboard. Remember to allow thinking time.
8. Introduce the criteria for determining the researchability and geographicality of the question and ask students to suggest which question is the most researchable for the purpose of the task.
9. Encourage the class to reach agreement on the inquiry question with reference to the criteria in step 8. This could be achieved by a vote or show of hands per question.
10. Once agreement is reached, you have completed the process of co-constructing a question with students where student voice is enabled and student engagement is quite high. The next part of the task is to conduct the bite-sized fieldwork activity to develop a response to the agreed-upon inquiry question

Conclusion

This chapter has detailed the place of inquiry in the Australian Curriculum: Geography, and the learning activities that together make up the parts of inquiry. It has explained a number of important reasons why inquiry is central in the teaching of this curriculum, and linked this to the work of educational theorists. The links between inquiry and thinking were described, and ways of creating a need to know outlined, with examples of how teachers have activated this in the classroom. Finally the process of creating inquiry questions was described with examples.

Bringing it together

1. How would you summarise each of the five groups of words in the sequence of inquiry detailed by the Australian Curriculum: Geography?

2. What are the main justifications for the use of inquiry given in this chapter?

3. List some of the main methods of ensuring that students develop thinking skills in your classroom.

4. Explain the use of 'looking through a window' in creating a need to know.

5. What are the key processes described in the example of co-creating an inquiry question?

References

Australian Curriculum Assessment and Reporting Authority [ACARA] (2016) *F–6/7 HASS, 7–10 History, 7–10 Geography, 7–10 Civics and Citizenship and 7–10 Economics and Business*, Version 8.3, Canberra: ACARA.

——(2020) *Australian Curriculum: Geography*, Canberra: ACARA, retrieved from www.australiancurriculum.edu.au/f-10-curriculum/humanities-and-social-sciences/geography/rationale

Bergsen, S., Meester, E., Kirschner, P. & Bosman, A. (2020) Constructivism is not a pedagogy, *David Didau*, 3 June, retrieved from https://learningspy.co.uk/literacy/constructivism-is-not-a-pedagogy

Caldis, S. (2015) Geography comes alive through fieldwork, *Geography Bulletin*, 47(1), 19–22.

——(2019a) STEM or HASS: Where is geography enabled or constrained?, *Professional Educator*, 1(20), 36–40.

——(2019b) Outstanding Educator in Residence review – bite-sized fieldwork: How fieldwork can play a bigger role in inquiry-based learning, in *Geography Education Matters: A publication by the Ministry of Education and Curriculum and Professional Development Division*, 2, retrieved from https://sites.google.com/moe.edu.sg/2019gem2/a-geographers-discourse

Caldis, S. & Kleeman, G. (2019) Geography and STEM, *Geographical Education*, 32, 5–10.

Firth, R. (2015) Constructing geographical knowledge, in G. Butt, ed., *MasterClass in geography education: Transforming teaching and learning*, London: Bloomsbury, pp. 83–101.

Rawling, E. (2007) *Planning your KS3 curriculum*, Sheffield: Geographical Association.

Roberts, M. (2003) *Learning through enquiry*, Sheffield: Geographical Association.

——(2013) *Geography through enquiry*, Sheffield: Geographical Association.

Spronken-Smith, R., Bullard, J., Ray, W., Roberts, C. & Keiffer, A. (2008) Where might sand dunes be on Mars? Engaging students through inquiry-based learning in geography, *Journal of Geography in Higher Education*, 32(1), 71–86.

Stemhagen, K., Reich, G. & Mutch, W. (2013) Disciplined judgement: Towards a reasonably constrained constructivism, *Journal of Curriculum and Pedagogy*, 10, 55–72.

What makes my geography lesson distinctive and powerful?

Susan Caldis

Learning objectives

By the end of this chapter, you should be able to:

- communicate an understanding about the nature and distinctiveness of geography and its pedagogies

- establish a response to the question 'What makes your geography lesson geographical?' and consider its enactment as part of individual practice

- develop an understanding about possible enabling and constraining influences on pedagogical practice in the geography classroom and propose a possible plan for action for responding to such influences

Introduction

As geography teachers, it is necessary to show students what makes geography distinctive, relevant and therefore powerful. The distinctiveness and relevance of a subject is shown through both content and pedagogy – pedagogical content knowledge, powerful knowledge, powerful pedagogy – bringing content knowledge to life for students through the way in which the subject is taught. Imagine a science lesson without experiments, a history lesson without site studies and source analysis of primary and secondary sources, an English lesson without interpretations of texts through imagery, metaphors and personification. Now imagine a geography lesson without fieldwork, or without the use of geographical tools such as maps and visual representations, or without the interpretation of information through the lens of place-based analysis, spatial reasoning and human–environment interconnections. Demonstrating an understanding of the subject – in this instance, geography – and demonstrating how to teach it is a requirement of Standard 2 in the Professional Knowledge domain of the Australian Professional

Standards for Teachers (AITSL 2017). Standard 2, 'Know the content and how to teach it', (AITSL 2017) becomes exemplified in a geography-specific context through the Professional Standards for the Accomplished Teaching of School Geography (Hutchinson & Kriewaldt 2010; Kriewaldt & Mulcahy 2010) and recent scholarship about powerful knowledge, geographical thinking and powerful pedagogies in geography (Brooks, Butt & Fargher 2017; Bustin 2019; Maude 2016, 2017; Roberts 2017; Young, 2008). The chapter explores the 'what', 'why' and 'how' of developing distinctive and powerful geography lessons through posing an overarching question for reflection: 'What makes your geography lesson geographical?' Throughout the chapter, the reader is challenged to reflect on and consider how they can continue to identify, maintain and build their pedagogical practice.

The 'what?' and 'why?' of a distinctive and powerful geographical geography lesson

To understand what a distinctive and powerful geographical geography lesson could be, it is important to develop a personal interpretation of geography. Difficulty in being able to concisely yet meaningfully define geography can lead to lack of clarity in the conceptual-isation, development and enactment of teaching, learning and assessment programs. It can therefore become difficult for students to connect with and embrace geography because the distinctiveness and relevance of the subject are not obvious. A meaningful understand-ing of the nature of geography has to occur for a teacher before it can be expressed adequately to students, and therefore enacted in teaching practice (Caldis 2015). Teacher subject identity is important to professional practice and differs from subject knowledge (Brooks 2016, 2021). Identifying as a 'geographer' and a 'geography teacher' assists teachers to make sense of geography as a subject in schools – it helps to bring the subject to life for students (Brooks 2016). Teacher subject identity informs decisions and actions upon pedagogical practice and professional learning opportunities, and therefore strengthens the distinctive presence of geography in classroom practice (Brooks 2021).

CONNECTION

From pitch to pedagogy, part 1

What makes geography geography?

It is inevitable that at least once in our career (if not each year!) we will be asked a question such as 'What is so important about geography?' or 'Why do we have to study geography?' It is up to us as geography teachers to have a resonating and meaningful answer to these serious questions.

Brainstorm three to five points to respond to the question 'What is so important about geography?' Consider the points and then rank them in order of importance to you as an aspiring geography teacher.

What makes a geography lesson geographical?

Brainstorm three to five points that confirm you have either been in or have taught a distinctly geographical geography lesson. Provide an outline for each point. Consider the points, then rank them in order of importance to you as an aspiring teacher of geography.

What is geography? Creating the pitch

'Geography – what is it about? And why do we have to study it?', 'How is geography different from history or science?' Questions such as these are often posed to teachers of geography by students and their parents or caregivers. Such questions can also be posed by colleagues from across the school. It is important for geography teachers to have a meaningful and targeted response.

In being able to respond to such a question, it is necessary for a geography educator to know why geography matters. Knowing why geography matters in terms of its real-life purpose and educational benefits is often reflective of the geography teacher's passion for the subject and evident in their professional practice (Brooks 2016, 2021; Geographical Association 2009). Murphy (2018, p. 83) notes that

> Geography, when taught well, has the capacity to challenge physical and mental bubbles that constrain thinking and experiences ... [Geography] can also enrich minds by fostering curiosity, awareness and appreciation for things we might otherwise take for granted.

While this quote may resonate with many, the way each teacher enacts and communicates key sentiments of the quote will differ. Our 'what matters' about geography, although grounded in similar theoretical understandings, needs to become a personalised story about what is important, interesting and powerful to us about this subject (Murphy 2018). From here, a pitch can develop to capture the attention of students and colleagues; to spark curiosity and wonder; and help you to 'walk-your-talk' as a geography teacher. It is a pitch that may strengthen and/or adjust over time.

A pitch can develop by looking through a window – often more than one window on more than one occasion. The windows include any combination of the following:

- theoretical windows such as empirical evidence, policy and curriculum documents
- practical windows such as lived experience, personal interest, engagement with networks, and time in the classroom teaching geography (Caldis 2020a, 2020b).

A pitch for geography needs to quickly present the subject and be attractive to others – the purpose of the pitch is to delight and convince. A pitch can be creative, formed as an argument, a problem or a question, invoke empathy, be a synthesis of varied contributors or be logically reasoned (Sebrae Entrepreneurship Education Reference Centre n.d.).

Pitching geography

The first GEOGstandard in the suite from the Professional Standards for the Accomplished Teaching of School Geography (Hutchinson & Kriewaldt 2010; Kriewaldt & Mulcahy 2010) reminds teachers to 'know Geography and the Geography curriculum'. Therefore, an important starting point is to re-communicate a formal definition of geography in an informal, succinct and interesting way. Chapter 1 of this textbook explores the nature of the subject. To refresh and build on learning from this chapter, it is now important to look at the 'what' of geography through the lens of 'why', and also to imagine and plan for the 'how' of teaching this subject.

Geography in schools is often declared to be the subject-vehicle through which students investigate and inquire about interactions between people and place to develop their global understanding and become equipped with the necessary attitudes and behaviours required to think about, devise and activate local interdisciplinary solutions to global problems (de Miguel Gonzales, Witham-Bednarz & DeMirci 2018; Maude 2010; Solem 2021). Such a position also relates to how geography as a discipline is defined in evidence-based policy such as the International Charter and Australia's strategic plan for the discipline. Definitions in policy contribute to the shaping of understanding about geography in curriculum documents:

> Geography in education is the study of the Earth, its natural and physical environments, human activities and social changes, the interrelationships and interactions of these and their effects from local to global scales; and, among many skills it uses mapping and fieldwork. (IGUCGE 2015, p. 1)

> Geography is a wide ranging and dynamic discipline that investigates many of the issues affecting the wellbeing of people and places in Australia and throughout the world ... geography brings an ability to integrate knowledge about the natural world, the social world and the humanities through its perspectives of space, place and environment ... [to find] answers to environmental, economic and social problems. (NCGS 2018, p. 1)

An excerpt from the Rationale for Geography in the Australian Curriculum positions geography as a subject where students can develop and question their global understanding, develop complex problem-solving skills and apply their skills and understanding to emerge as active and informed citizens:

> Geography inspires curiosity and wonder about the diversity of the world's places, peoples, cultures and environments. Through a structured way of

exploring, analysing and understanding the characteristics of the places that make up our world, Geography enables students to question why the world is the way it is, and reflect on their relationships with and responsibilities for that world. (ACARA 2020)

A synergy exists between the definition of geography as a discipline and as a school subject. Curiosity, investigation and action are emphasised through a conceptual, multi-disciplinary and multiperspective lens. Such emphasis signals that the skills of application, problem-solving, data-gathering and analysis will be important for geographical learning. Therefore, the way geography is experienced and learned will be distinctive. Consequently, geography as a school subject sets the foundation for the future of the discipline. The future of the discipline spans from being a core and elective learning subject with viable candidature to being identified as a relevant, rigorous area of research in higher education, and also as a diverse and interesting career pathway where significant contributions to society can occur from the personal to the global scale.

CONNECTION

From pitch to pedagogy, part 2

What makes geography geography?

Revisit the pitch-points for this question from the previous Connection box, 'From pitch to pedagogy, part 1'.

In response to the question 'What is so important about geography?', do the three to five points become strengthened, adjusted or changed as a result of reading the chapter so far? Are key ideas from a theoretical window(s) evident? Does a practical window feature in one or more of your points?

Pedagogical content knowledge + powerful knowledge = powerful pedagogy

Content knowledge alone is not sufficient for the effective teaching of geography (Brooks 2011; Mitchell 2018). As a result, the development of pedagogical content knowledge (PCK), reflection on PCK and developing an understanding about the powerful knowledge of geography are required by the teacher to bring relevance and meaning to the classroom (Walshe, Driver & Keenoy 2021). PCK surrounds the multiple specific strategies and activities for promoting student understanding by representing and organising subject matter so it becomes accessible to students (Lane 2015; Shulman 1986). Powerful knowledge is about developing new ways of thinking within a subject. For geography, this typically will occur through exploring the conceptual underpinnings to interpret and act upon complex issues in a subject-specific, analytical, systematic and confident way (Fögele 2017; Maude 2017; Young 2008). In doing this, teachers will be able to demonstrate achievement of the Australian Professional Standards for Teachers Standard 2, 'Know the content and how to teach it' (AITSL 2017).

Pedagogical content knowledge

PCK refers to knowledge about the 'multiple methods of representing and organising subject content to make it more easily comprehensible to students' (Lane 2015, p. 29). Overall, teacher use of PCK strategies will assist with developing student understanding about a subject because the more a teacher focuses on their subject-specific pedagogical decision-making processes, the more deeply they start to understand the subject and how to transform the subject into a teaching-appropriate form (Lane 2009, 2015; Shulman 1986; Walshe, Driver & Keenoy 2021).

> Shulman (1986) identifies four key areas of knowledge that teachers develop as they progress from college graduates to accomplished professionals: knowledge of the key concepts and structure of a teaching subject (content knowledge); knowledge of relevant curriculum documents and supporting resources (curriculum knowledge); general pedagogical knowledge; and pedagogical content knowledge (PCK). (Lane 2009, p. 40)

The multiple methods of representing and organising subject content in a distinctive way through PCK strategies will enable a teacher to consider the issue of how best to teach the subject (Walshe, Driver & Keenoy 2021). In so doing, the teacher should be able to choose suitable analogies, examples or applications of a concept or skill to make the content accessible and interesting to a particular group of students (Lane 2009, 2015). From here, an entry point into powerful knowledge and powerful pedagogy occurs.

Powerful knowledge

Powerful knowledge emerges from the field of sociology, and is understood to comprise reliable explanations for, and new ways of thinking about, the natural and social worlds – to predict, to explain, to engage in debates of significance, to envisage alternatives and to think beyond individual and personal experiences (Young 2008, 2014). Powerful knowledge is about developing new ways of thinking, typically through conceptual underpinnings, to interpret and act upon complex issues in a subject-specific, analytical, systematic and confident way (Fögele 2017; Young 2008). In geography, powerful disciplinary knowledge needs to be positioned alongside everyday practical knowledge to enable students to enhance their capabilities (Solem 2021) by:

- developing new ways of thinking about the world
- developing power, independence and ownership over their interpretations, thoughts and development of knowledge
- better explaining, analysing and understanding the world

- following and participating in current and significant debates about local, national and global issues

- thinking about alternative futures and how to influence them (Maude 2015, 2016, 2017).

It is not about developing a list of content to be taught, but rather developing the capacity to think geographically (Bustin 2019; Maude 2016). Scholarship within geographical education therefore proposes that four of the subject's key concepts – place, space, environment, interconnections – will enhance geographical thinking processes. Geographical thinking through the core concepts will elicit a powerful and distinctive knowledge of geography because the focus of interpretation and meaning-making of content is distinctly geographical (Bustin 2019; Fögele 2017; Maude 2017). As a result, powerful knowledge in geography helps teachers and students to discern and justify responses to the question 'Where is the geography?' (Bustin 2019) and to understand what makes a geography lesson geographical.

Powerful pedagogy

When a teacher can pitch their personalised understanding of geography and articulate a rationale for their selection of subject-specific pedagogies to assist students with their development of geographical thinking skills and powerful knowledge, the teacher is equipped to enact powerful pedagogical practice. The identification, selection and enactment of powerful pedagogical strategies therefore arise from an understanding of the powerful knowledge of geography and its evidence in the curriculum. Once such powerful knowledge is determined, a teacher can focus on the important question of the 'why' and 'how' of what to teach (Bustin 2019). In geography, powerful pedagogy should be part of classroom culture to promote critical geographical thinking and encourage the active involvement of students in their learning through investigation and fieldwork to shape dialogue and construction of their geographical knowledge and understanding (Roberts 2017). Powerful pedagogy in geography occurs by:

- making connections between students' everyday knowledge and school geography knowledge

- transforming the way students interpret and understand the world through developing geographical thinking processes and encouraging active investigation

- developing awareness about the values dimension of decisions affecting environments and communities at local, national and global scales

- developing geographical skills such as map reading and spatial data interpretation to gather, analyse and communicate information

- encouraging active learning through posing questions, using geographical tools and dialogic critical thinking processes (Roberts 2017, 2021).

The ideas of powerful pedagogy are evident in the Professional Standards for the Accomplished Teaching of School Geography (Hutchinson & Kriewaldt 2010; Kriewaldt & Mulcahy 2010), also known as the GEOGstandards. As mentioned in previous chapters, the GEOGstandards are nine evidence-based standards, identified in Table 8.1, of exemplary pedagogical practice in geography as demonstrated by experienced specialist geography teachers across Australia. The GEOGstandards most relevant to pedagogical content knowledge, powerful knowledge and powerful pedagogy are bolded in Table 8.1.

Table 8.1 The Professional Standards for the Accomplished Teaching of School Geography

Standard	Overview
1. Knowing geography and the geography curriculum	As the teacher: understand the discipline including concepts and skills; understand the curriculum; understand geography draws from the social sciences, physical sciences and humanities; and make connections with other curricula and learning areas.
2. Fostering geographical inquiry and fieldwork	Allow students to carry out a range of structured and open-ended inquiries; and undertake inquiry in the field, selecting and using geographical tools.
3. Developing geographical thinking and communication	Encourage and support student understanding about spatial reasoning; conceptual interdependencies, interconnections and assemblages; real-world contexts at a range of scales; and lived experience as a personal geography.
4. Understanding students and their communities	Use local community contexts and personal geographies to connect, enhance and enrich conceptual and perspective-focused learning.
5. Establishing a safe, supportive and intellectually challenging learning environment	Facilitate students becoming active participants in their own learning by creating a need to know and creating conditions for students to question complex geographical ideas.

Table 8.1 (*cont.*)

Standard	Overview
6. Understanding geography teaching – pedagogical practices	Teachers: have extensive understanding about pedagogical content knowledge; encourage students to gather information from a variety of sources; use fieldwork; introduce a range of tools to students.
7. Planning, assessing and reporting	Plan, monitor and assess geographical learning through a range of formal and informal methods; recognise achievement and provide direction for improvement; use diagnostic assessment to inform teaching practice.
8. Progressing professional growth and development	Engage with professional learning communities; recognise geography is an evolving subject requiring regular updating of content knowledge
9. Learning and working collegially	Actively engage with professional community; share expertise; build a culture of professional improvement; promote geographical education

Source: Hutchinson & Kriewaldt (2010); Kriewaldt & Mulcahy (2010); GEOGstandards (2010).

CONNECTION

From pitch to pedagogy, part 3

What makes geography geography?

Revisit the pitch-points for this question from the previous Connection boxes.

In response to the question 'What is so important about geography?', do the three to five points become strengthened, adjusted or changed as a result of reading the chapter so far? What would you add or change to this pitch to provide a presence to pedagogy?

What makes a geography lesson geographical?

Revisit the points for this question from the first Connection box, 'From pitch to pedagogy, part 1'.

In response to the question 'What makes a geography lesson geographical?', do the three to five points become strengthened, adjusted or changed as a result of reading the chapter so far? Does the order of importance change? Once the three to five points have been confirmed, identify the associated powerful pedagogy/ies and GEOGstandard(s).

The 'how?' of a distinctive and powerful geographical geography lesson

The 'how' of developing a distinctive and powerful geography lesson requires reflexive practice – being able to discern, deliberate and dedicate action in response to the most influential enabling and constraining factors that impact teaching practice. Use of a protocol can sometimes assist with the act of curriculum-making and reflexivity. The final part of the 'how' is planning for action, and typically this occurs through the development of SMART goals – which are specific, measurable, achievable, realistic and time-based or trackable. The 'how' of developing a distinctive and powerful geography lesson also requires drawing on understanding about teacher subject identity (Brooks 2016, 2021), pitch (Caldis 2015, 2020a, 2020b), pedagogical content knowledge strategies (Lane 2009; Shulman 1986; Walshe, Driver & Keenoy 2021) and powerful knowledge (Maude 2016, 2017) when selecting appropriate powerful pedagogies as part of the 'curriculum-making' process (Lambert & Morgan 2010).

Curriculum-making

Curriculum-making requires the involvement of both teacher and student in the holistic conceptualisation of what, why and how learning occurs in a course or subject; it is a long-term process beyond the immediate day-to-day lesson planning to identify the core learning required from the unit of work (Bustin 2019). Curriculum-making is referred to as the creative act of interpreting a curriculum and developing a coherent, challenging, engaging and enjoyable unit of work (Lambert & Morgan 2010). Overall, the curriculum-making process involves considered planning to enact authentic links between curriculum understanding, content knowledge, pedagogical practice, local context and student experience (personal geographies). The curriculum-making process is explored further in Chapter 11.

Reflexive practice

A reflexive practitioner is one who is self-monitoring, self-aware and continually deciphering self (you as the teacher), circumstance (events occurring) and phenomenon of interest (e.g. enacting a geographical geography lesson) against the context of enabling or constraining influences arising from personal, structural and cultural emergent properties (Archer 2010). The emergent properties will encourage (enable) or limit (constrain) teaching practice; once they are discerned (identified) and deliberated (thought about), a teacher should determine which emergent property is the most influential on practice. Personal emergent properties relate to personal values and beliefs, such as a conviction to emphasise the enactment of student-centred active learning approaches. Structural emergent properties relate to processes, structures and policies already in place, such as syllabus documents and timetables. Cultural emergent properties relate to the culture of place and context. After the discernment and deliberation process, a plan for action needs to occur – dedication (Archer 2010). The action needs to focus on the influential emergent properties that can be made to either maximise the enabling influence or mitigate the influence of constraint. It is the taking of action that separates reflexivity from reflection (Figure 8.1).

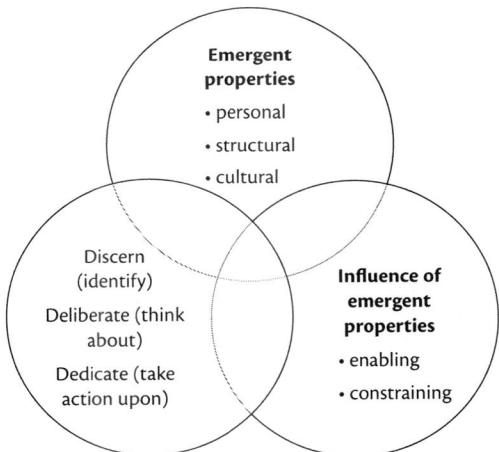

Figure 8.1 Understanding reflexive practice

CONNECTION

Enacting a distinctive and powerful geography lesson, part 1

Understanding the influences on my pedagogical practice

Reflect on a geography lesson from your most recent professional experience placement. Complete Table 8.2 by considering the distinctive and powerful features of your geography lesson together with the enabling and/or constraining influences upon your practice.

Table 8.2 Influences on pedagogical practices

Emergent property (there may be more than one item for each property)	Enabling or constraining	Tick the corresponding box to show *the most* influential property(ies) on your teaching practice (enabling and constraining)	Identify an action to maximise the enabling influence and an action to mitigate the constraining influence
Personal			
Structural			
Cultural			

A protocol for guiding personal reflection and/or peer feedback

Use of a protocol is beneficial when developing or reflecting on the distinctive, relevant and powerful nature of a geographical geography lesson. A protocol can assist with identifying and contemplating areas of existing practice with a view to understanding 'what now' and what actions might be most appropriate for 'what next'.

When thinking about what makes a geography lesson geographical, the features in Figure 8.2 may be helpful in preparing for a more detailed reflection (Caldis 2019).

Figure 8.3 provides an excerpt from a protocol for guiding the creation of a distinctive, relevant and powerful geographical geography lesson (Caldis 2021) (the full protocol is available on the website for this text). The protocol can be used for reviewing a written and/or enacted lesson plan as part of personal reflection. The protocol can also be used to scaffold feedback and feedforward as part of peer assessment. It is not intended for each element of the protocol to be evident in each geography lesson.

As a result of participating in a distinctive and powerful geography lesson your students become geographers who learn to both think and communicate in a geographically distinctive way. In reflecting upon the full protocol in the online resources, you are invited to first consider what your next steps will be to ensure your teaching practice in the geography classroom becomes as distinctive, relevant and powerful as possible. Second, it is worth pondering the implications for the development of future assessment tasks (Caldis 2019). What becomes emphasised in our pedagogy as a geography teacher should also become evident and valued in our geography assessments. (This is explored further in Chapter 11.)

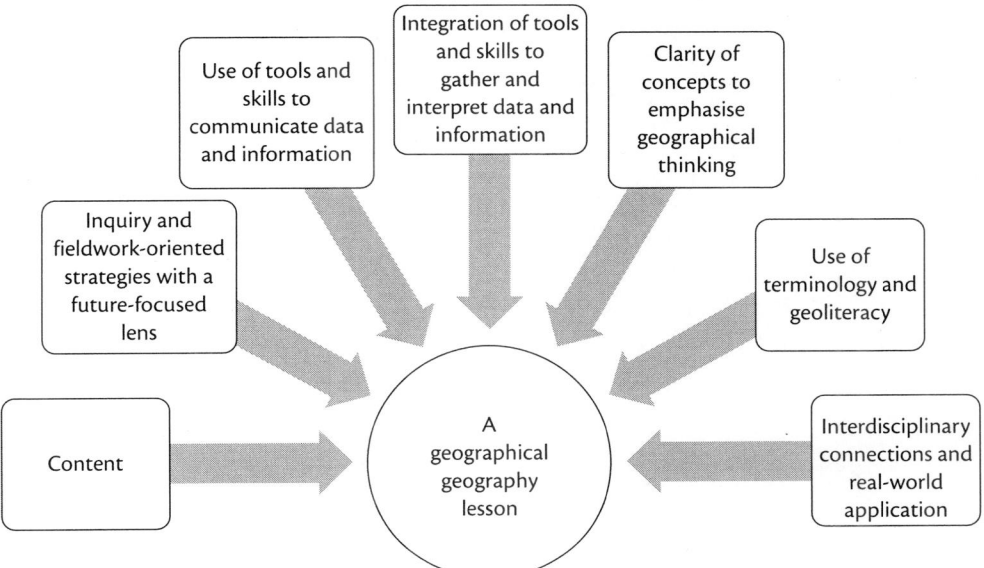

Figure 8.2 What makes a geography lesson geographical?

Source: Caldis (2019).

CONNECTION

Developing and enacting a distinctive and powerful geography lesson, part 2

Where is the geography in my geography lesson?

Reflect on a geography lesson from your most recent professional experience placement.
  With reference to the protocol, which areas would you like to emphasise to ensure the unit of work and therefore as many lessons as possible become distinctive, relevant and powerful geographically?

Conclusion

The creation of distinctive and powerful geography lessons helps to maintain the relevance of geography. Harnessing the powerful knowledge and powerful pedagogies of geography will help students to think and interpret processes, context and lived experiences in a geographical way. The development of geographical geography lessons occurs by:

- asking questions and thinking about possible responses to progress beyond factual knowledge towards higher levels of thinking (Maude 2017)

Date	Time
Lesson duration	Unit of work / topic
Year Level	

Rating scale: 1 – 5 where 1 = no use, 3 = some use, 5 = extensive use	
Rating scale: M = implicit use, X = explicit use	

Posing or generating inquiry questions	Please circle: 1 2 3 4 5 M X Comment:
Explicitly using the concepts	Please circle: 1 2 3 4 5 M X Comment:
Designing and/or enacting fieldwork	Please circle: 1 2 3 4 5 M X Comment for:
Articulating and/or enacting interdisciplinary connections	Please circle: 1 2 3 4 5 M X Comment:
Using geospatial technologies to harness and/or communicate geographical data and information	Please circle: 1 2 3 4 5 M X Comment:

Figure 8.3 An excerpt from a protocol to guide the development of a distinctive and powerful geography lesson

Source: Caldis (2021).

- encouraging students to gather data in response to questions through fieldwork and secondary research methods (Caldis, 2015)
- considering 'the view from the window' in the context of key geographical concepts such as place, space, environment, interconnections and sustainability,

and how they relate to each other to make links, comparisons and generalisations (Caldis 2020b; Lambert 2017)

- applying knowledge and conceptual understanding about global geographical issues and problems to a local context – using a local context to make sense of the global (Martins 2017). The synthesis of ideas, thinking and discussion about an alternative future together with a way to influence such a future can then occur (Maude 2017).

Bringing it together

1. Why is it important to develop a pitch for geography?

2. What are some pedagogically distinctive features of a geography lesson?

3. How can powerful knowledge and powerful pedagogy be utilised in the development of a lesson or teaching program for geography?

4. Why is reflexivity theory useful in helping a teacher improve their pedagogical practice in the geography classroom?

5. What is it to you that makes a geography lesson geographical?

References

Archer, M.S. (2010) Routine, reflexivity and realism, *Sociological Theory*, 28(3), 272–303.

Australian Curriculum Assessment and Reporting Authority [ACARA] (2020) *Foundation to Year 10 Australian Curriculum: Geography*, Canberra: ACARA, retrieved from www.australiancurriculum.edu.au/f-10-curriculum/humanities-and-social-sciences/geography/rationale/

Australian Institute for Teaching and School Leadership [AITSL] (2017) *Australian Professional Standards for Teachers*, Canberra: Education Services Australia.

Brooks, C. (2011) Geographical knowledge and professional development, in G. Butt, ed., *Geography education and the future*, London: Continuum, pp. 165–80.

——(2016) *Teacher subject identity in professional practice*, London: Routledge.

——(2021) Teacher identity, professional practice and online spaces, in N. Walshe & G. Healy, eds, *Geography education in the digital world*, London: Routledge, pp. 7–16.

Brooks, C., Butt, G. & Fargher, M. (2017) *The power of geographical thinking: International perspectives on geographical education*, Dordrecht: Springer.

Bustin, R. (2019) *Geography education's potential and the capability approach: GeoCapabilities and schools*, London: Palgrave Macmillan.

Caldis, S. (2015) Geography comes alive through fieldwork, *Geography Bulletin*, 47(1), 19–22.

——(2019) What makes your geography assessment geographical?, *Geography Bulletin*, 51(1), 67–70.

——(2020a) A geography lesson out of every window: Fieldwork in 50 lessons from Singapore, *Geography Bulletin*, 4, 76–82.

——(2020b) Pitch and practice: A geography lesson out of every window, *GEOBUZZ 2020, Journal of the Geography Teachers' Association of Singapore*, retrieved from www.gtansw.org.au/wp-content/uploads/2020/10/07_GTA-Bulletin-Issue-4-2020_ Fieldwork-in-50.pdf

——(2021) Transition into the profession and transformation of pedagogical practice, unpublished thesis.

de Miguel Gonzales, R., Witham-Bednarz, S. & DeMirci, R. (2018) Why geography education matters for global understanding, in A. DeMerci, R. de Miguel Gonzalez & S. Witham-Bednarz, eds, *Geography education for global understanding*, Dordrecht: Springer, pp. 15–28.

Fögele, J. (2017) Acquiring powerful thinking through geographical key concepts, in C. Brooks, G. Butt. & M. Fargher, eds, *The power of geographical thinking*, Dordrecht: Springer, pp. 59–74.

Geographical Association (2009) *A different view: A manifesto from the Geographical Association*, London: Kingsbury Press.

GEOGstandards (2010) *Professional Standards for Accomplished Teaching of School Geography*, retrieved from www.agta.asn.au/files/Professional%20Standards/ geogstandards.pdf

Hutchinson, N. & Kriewaldt, J. (2010) Developing geography standards: Articulating the complexity of accomplished geography teaching, *Geographical Education*, 23, 32–40.

International Geographical Union Commission on Geographical Education [IGUCGE] (2015) *International research in geographical education*, retrieved from www.igu-cge .org/wp-content/uploads/2018/02/International-Declaration-on-Research-in-Geography-Education-FULL-DOCUMENT-JUNE-2015.pdf

Kriewaldt, J. & Mulcahy, D. (2010) Professional standards for teaching school geography, *Curriculum & Leadership Journal*, 8(20), retrieved from www.curriculum.edu.au/leader/ professional_standards_for_teaching_school_geograp,31904.html?issueID=12165

Lambert, D. (2017) Thinking geographically, in M. Jones, ed., *The handbook of secondary geography*, Sheffield: Geographical Association, pp. 20–9.

Lambert, D. & Morgan, J. (2010) *Teaching geography 11–18: A conceptual approach*, Oxford: Oxford University Press.

Lane, R. (2009) Articulating the pedagogical content knowledge of accomplished geography teachers, *Geographical Education*, 22, 40–50.

——(2015) Experienced geography teachers' PCK of students' ideas and beliefs about learning and teaching, *International Research in Geographical and Environmental Education*, 24(1), 43–57.

Martins, F. (2017) Teaching to develop geographical thinking, in C. Brooks, G. Butt & M. Fargher, eds, *The power of geographical thinking: International perspectives on geographical education*, Dordrecht: Springer, pp. 199–209.

Maude, A. (2010) What does geography contribute to the education of young Australians? *Geographical Education*, 23, 14–22.

——(2015) What is powerful knowledge and can it be found in the Australian Curriculum: Geography?, *Geographical Education*, 28, 18–26.

——(2016) What might powerful geographical knowledge look like?, *Geography*, 101(2), 70–6.

——(2017) Applying the concept of powerful knowledge to school geography, in C. Brooks, G. Butt & M. Fargher, eds, *The power of geographical thinking: International perspectives on geographical education*, Dordrecht: Springer, pp. 27–40.

Mitchell, J.T. (2018) Pre-service teachers learn to teach geography: A suggested course model, *Journal of Geography in Higher Education*, 42(2), 238–60.

Murphy, A.B. (2018) *Geography: Why it matters*, Bristol: Polity Press.

National Committee for Geographical Sciences [NCGS] (2018) *Geography: Shaping Australia's future*, Canberra: Australian Academy of Science.

Roberts, M. (2017) Geographical education is powerful if . . ., *Teaching Geography*, 42(1), 6–9.

——(2021) Why is powerful knowledge not enough? The power depends on powerful pedagogy, keynote lecture for IGU-CGE Future Teachers for the Planet e-symposium, retrieved from www.geographie.hu-berlin.de/de/abteilungen/ didaktik/ftftp/keynotes

Sebrae Entrepreneurship Education Reference Centre (n.d.) What is a pitch and how to use it in education?, retrieved from https://cer.sebrae.com.br/what-is-a-pitch-and-how-to-use-it-in-education/?lang=en

Shulman, L. (1986) Those who understand: Knowledge growth in teaching, *Educational Researcher*, 15(2), 414.

Solem, M. (2021) Powerful geography, keynote lecture for IGU-CGE Future Teachers for the Planet e-symposium, retrieved from www.geographie.hu-berlin.de/de/abteilungen/ didaktik/ftftp/keynotes

Walshe, N., Driver, P. & Keenoy, M.-J. (2021) Navigating the theory–practice divide, in N. Walshe & G. Healy, eds, *Geography education in the digital world: Linking theory and practice*, London: Routledge, pp. 26–37.

Young, M. (2008) From constructivism to realism in the sociology of the curriculum, *Review of Research in Education*, 32, 1–32.

——(2014) Powerful knowledge as a curriculum principle, in M. Young, D. Lambert, C. Roberts & M. Roberts (eds), *Knowledge and the future school: Curriculum and social justice*, London: Bloomsbury, pp. 65–88.

CHAPTER

9

Fieldwork

Stephen Cranby

Learning objectives

By the end of this chapter, you will be able to:

- identify the centrality of fieldwork within the discipline of geography
- understand different fieldwork pedagogy, and its contribution to students learning
- develop skills and knowledge in relation to the practice and implementation of fieldwork within the school context

Introduction

Geographers have always had a love affair with fieldwork. It has been their staple since the beginning. As Kalyani (2012, p. 303) succinctly puts it, 'fieldwork and geography have essentially been thought of by most Geographers, as coexisting, complementary, and an indivisible media for conveying a comprehensive understanding of the environment'. In his book *The Geographer's Art*, Haggett (1995, p. 31) takes it back further, describing how the earliest explorers, collectively through their contribution of 'fragmentary evidence, sketch maps and field books, logs and travel diaries, bearings and distances', contributed to our 'rudimentary knowledge of [how] the vast spaces of the world was made up'. On a more down-to-earth scale, Winchester (2001) describes the life of William Smith, a blacksmith's son from Oxfordshire who spent 20 years criss-crossing England, investigating, identifying and mapping the different rock strata he saw beneath his feet. Eventually (in 1815), he produced a map describing the geology of Britain. They really were the first 'geographers' to use fieldwork to make sense of their world.

Through the work of early explorers, individuals such as William Smith and intellectual societies such as the Royal Geographical Society, the discipline of geography evolved with fieldwork as one of its central tenets. In the school context, however, 'fieldwork, though central to geography teaching since the discipline first evolved . . . always struggled under the the pressures of safety, cost and the internal managerial and curriculum pressures in schools' (Foskett 1997, p. 200). Such pressures have always existed, and continue to exist, in schools; they can lead to the dimunition of the value of fieldwork as part of the geography curriculum. However, fieldwork's value has always been recognised, and 'with effective planning and management and a commitment to the educational and personal benefits to pupils of fieldwork, geography teachers can ensure that it remains as one of the most significant learning experiences that pupils have during their school career' (Foskett 1997, p. 200).

In a somewhat prescient view, Haggett (1995, p. 31) notes that 'the choice is not between the traveller-explorer of an early century and a computer-tied geographer of the late twentieth century, but rather how we are to bridge the gap between local field observation and scientific generalisations that hold over a wider spatial domain'. Here he provides a portent of the potential quandary between on-the-ground fieldwork and the availability of satellite imagery, live cams, smartphones and the all-pervasive internet – the fieldwork of the future?

Centrality of fieldwork

Geography teachers have a collective understanding of and consensus about what fieldwork is and its role within their geography teaching. In its simplest form, fieldwork helps to make sense of what students learn in the classroom, but it is much more than that. Teachers unanimously affirm its central role in providing an extension to classroom learning by involving students in active data collection in the field. The data collected are brought back to the classroom for further processing, analysis, evaluation and synthesis, building upon students' earlier classroom experience. This emphasises the opportunity provided by fieldwork to reinforce students' classroom learning through practical application of theory, inquiry-based learning and experiential learning (Cranby 2002).

Practising teachers note that some of the greatest discussions, as well as subsequent learning exhibited by students, take place while in the field. However, these do not just occur in the field. As Huckle (1983, p. 62) observes, 'Such discussions in the field lead to similar debates in the classroom'. These occur through the process of data analysis and presentation, evaluation and synthesis, as students review and assess their fieldwork results. Kalyani (2012, p. 304) aptly sums up the role of fieldwork in geography teaching by noting that 'fieldwork cannot be taken as a complementary part of geographic learning, but it should be taken as an integral component'.

Feedback from practising teachers over numerous fieldwork workshops (Cranby n.d.) has consistently identified the following characteristics that broaden our understanding of fieldwork:

- inquiry-based learning, emphasising learning rather than teaching, through student observation, data collection and experiencing the environment

- hands-on learning, where students learn and apply skills beyond those used in the classroom

- experiential learning in action, reinforcing and extending what students have learned in the classroom and putting theory into practice

- a richer and more diverse experience than that of the classroom – more interactive and rewarding.

In the school context, it is important to clarify for students the distinction between fieldwork undertaken in geography and excursions carried out elsewhere in the curriculum. As outlined above, fieldwork is active and engaged learning, compared with excursions, which tend to encourage passive learning. The importance of generating a 'culture' of fieldwork (to develop independence in senior students) cannot be underestimated. The earlier students are introduced to the expectations of fieldwork, the more effective future fieldwork activities will be, resulting in development of their skills, independence and enjoyment of the activity.

Short-answer question

Where and to what extent is fieldwork identified directly or indirectly across Years 7–12 in the Australian Curriculum: Geography?

Benefits of fieldwork

Fieldwork is more than just an extension of classroom content and skills, and a different learning environment. It contributes to students' understanding and application of key geographic concepts such as location, distribution, association, interaction, movement and change. It has a wider impact on student learning in general through its positive impact on long-term memory, individual growth, social skills, development of affective and cognitive thinking, and providing opportunities for higher order learning (Laws 1984; Oost, De Vries & Van der Schee 2011).

Some affective benefits arising from students' fieldwork experiences include:

- independence, ownership, and responsibility for collecting their own data

- cooperative learning and social experience of working with peers

- opportunity to challenge perceptions, stereotypes or value systems, and to view common sites in a different (more critical) way

- greater appreciation of, and interaction with, the environment

- opportunity to be more engaged and invigorated, leading to greater satisfaction.

Huckle (1983, p. 61) aptly synthesises the benefits of fieldwork in geography when he points out that 'what is vital to fieldwork is the hope that something rubs off, that there is an experience beyond that gained by using . . . secondary sources'.

Students' fieldwork experiences stay with them forever. Former students of the author, reflecting on the long-lasting nature of their fieldwork experiences, have commented that it 'allows for a far deeper appreciation and understanding of the geographical processes bring explored', 'made me be more conscious of the environment and appreciate how everything interacts with each other' and instilled 'an everlasting interest and respect for the natural world'. The fact that many ex-students work either directly or indirectly in fields that inform or are informed by geography, and others use their experience in their daily lives, is a testament to the impact that fieldwork, as part of geography teaching, can have on student outcomes.

Pause and think

Perception, attitude, and value – prepare us, first of all, to understand ourselves. Without self-understanding we cannot hope for enduring solutions to environmental problems, which are fundamentally human problems. (Tuan 1974)

How can fieldwork contribute to the achievement of this quote from Yi-Fu Tuan?

Pedagogy of fieldwork

Teachers reflecting on how fieldwork benefits their students tend to identify two distinct areas of benefit (see Table 9.1). The first is fieldwork contributing to students' 'internal' growth beyond the specific context of the fieldwork – to their wider geographic appreciation. The second is its contribution to students' 'external' growth and development of learning skills and behaviours, beyond those of the activity itself.

In considering what fieldwork brings to their teaching (see Table 9.2), teachers value the opportunities to develop the student–teacher relationship beyond that in the classroom, through the collaborative nature of fieldwork and different teacher–student interactions. Importantly, fieldwork enables teachers to refresh their enthusiasm for geography and to share this with their students. For students, the link between classroom teaching and the outside world attracts them, while the challenges and personal opportunities for learning generate enthusiasm for the subject.

Fieldwork is not an 'add on' to illuminate classroom teaching, but a part of the curriculum that requires careful and thorough planning. A one-day fieldwork activity deserves the equivalent effort and planning of four to six lessons. As Foskett (1997, p. 195) states, all fieldwork activities require good lesson planning, 'with the need for

Table 9.1 How fieldwork benefits students

Benefits for students	
A student's 'internal' growth	**A student's 'external' growth**
• Experience of unfamiliar environments • Development and awareness of the big picture (world) • Appreciation and understanding of the interaction of elements within environments • Challenging stereotypes, perceptions or value systems • Expansion of their understanding and interest in the real world	• Development of responsibility and ownership through the collection of data • Providing an opportunity to be 'engaged' • A chance for those whose learning is stronger outside the classroom to 'strut their stuff' • An out of classroom experience, leading to development of different skills

Source: Cranby (2002).

Table 9.2 What does fieldwork bring to our teaching?

	For students	For teachers
Enhances the subject (e.g. attractive to students)	Provides clarification and relevance to what occurs in the classroom, a practical application and connection to the outside world.	Provides a focus for classroom teaching. The practical skills and real-world data of fieldwork bring a new reality to the classroom.
Enthusiasm	Caters for a wide range of learning styles, provides challenges and is more personal.	Offers opportunities the classroom cannot; provides a more varied relationship and rapport with students, a change in the variety of teaching strategies used, and can revitalise and reinvigorate teachers.

Source: Cranby (n.d.).

clarity of aims, objectives and intended learning outcomes, the selection of appropriate methods (either for or with the students), the use of suitable resources and equipment, and careful attention to issues such as timing and sequencing of activities'.

As important as it is to carefully plan any fieldwork activity, it also needs to be integrated into classroom learning. When challenged to identify the links between geography and fieldwork, teachers are quite adamant, and universally agree that it is 'integral' and 'essential'. The link between fieldwork and classroom practice, and its importance for developing a 'sense of place' in students, are paramount (Cranby 2002).

Cooperative learning

Fieldwork offers the ultimate cooperative learning exercise. Through the planning and design of fieldwork tasks, opportunities for cooperative learning are maximised. While the interpretation and communication of the fieldwork results may be the product of an individual, the cooperative learning processes prior to this occur through students working together on shared tasks, collecting data as individuals or small groups, and sharing results. Students and teachers discuss their observations and insights during the data-collection process. On extended fieldwork trips, some of the most valuable learning takes place in the discussions around the fireplace in the evenings: 'It all comes down to the Socratic question of "Why?" and then the pursuit of knowledge and the sharing of that knowledge with a community of active learners' (Westhorpe 2011, p. 15).

While the curriculum sets the focus of fieldwork, there are many opportunities to involve students in negotiating the design of the task. With negotiation, students have some say in setting the directions of the work undertaken. This empowerment encourages student motivation and interest, resulting in an improved commitment to the task at hand. Furthermore, 'it is in the nature of the negotiation process, the development of the topic by the student, that the role of the teacher changes to that of being a true "facilitator" of learning rather than the more dogmatic passer-on of information' (Cranby 1998, p. 24). Through negotiation, students are more able to select their own tasks or pathways through the work, enabling discernment in learning to take place more easily. However, not all students flower in the negotiation process, and some may require a degree of structure or a fallback position.

The role of the teacher in the learning process changes from planning, resourcing and embedding the fieldwork in the curriculum (areas of study, inquiry sequences and outcomes), which any negotiated fieldwork activity must meet, to the fieldwork itself, where as a facilitator they are on the go constantly, taking the opportunity to spend lengths of time with individuals and cater for individual differences.

In practice, what can be negotiated? It all depends on the age or year level of the students and their experience with fieldwork. Students can be involved in the design of:

- inquiry question(s) and development of hypothesis
- methodology – fieldwork techniques and data collection

- choice of presentation formats
- assessment processes (mainly in junior and middle levels).

Fieldwork design

Geography fieldwork 'should be compatible with the educational experience and state of progress of the students' and should be 'integrated into the structure and learning objectives' of the teaching and learning program (Kent, Gilbertson & Hunt 1997, p. 319). This involves the development of an appropriate fieldwork inquiry sequence that includes identifying a problem to investigate, developing a hypothesis to examine, collection of primary data, presentation of results, including their description, analysis, and interpretation, evaluation of the hypothesis and conclusions (Boardman 1983, p. 120). Similarly, the design of geography fieldwork should acknowledge the sequential development of skills in, and use of, appropriate fieldwork techniques in students' progress (see Chapter 5). As Foskett (1997, p. 191) emphasises, 'the importance of skill development, data collection and hypothesis testing' cannot be under-estimated.

While the curriculum identifies the type of site or topic of fieldwork appropriate for each level, it is important to consider the progression of students through the school's program over the years. It should show a progression or increase in distance, duration and complexity. Table 9.3 outlines the progression of a student's fieldwork experience in terms of a site's distance from the school, and the duration of time spent there. There is a clear spatial/temporal association between the two: the greater the travel time, the more time spent at the site.

Table 9.3 Fieldwork progression – distance and duration

Distance	Duration
School grounds	Within a lesson
School grounds	Within lesson or part day
Local region	Part day or whole day
Wider region	Whole day
Distant locality	Residential (few days/week)
Interstate	Residential (week)
Overseas	Residential (at least one week)

Source: Adapted from Foskett (1997, p. 192).

This association has implications for planning fieldwork for different age groups/ year levels. Again, the concept of progression comes into play. As students' experience of fieldwork grows, so does their ability to move out and spend more time on the task. Generally, the younger the student, the more school and locally based brief fieldwork experiences. As students get older and more experienced, more distant fieldwork sites with longer durations become possible.

Pause and think

1. Consider what you could do with the following list of broad fieldwork topics in terms of the Australian Curriculum: Geography units and year levels. Place these topics in Table 9.4 according to the year level and distance from school that the fieldwork you envisage could occur. Share and discuss your 'placements' with other colleagues.

    — Bushfires

    — Chemical wastes

    — Litter

    — Market gardens

    — National parks

    — Public transport

    — Recycling

    — River catchment

    — Shopping

    — Solar energy

    — Tourism

    — Traffic patterns

    — Urban waterways

    — Migration

    — Recreation

Table 9.4 Fieldwork topic: Distance by year level

Distance	Year level		
	Junior	Middle	Senior
School grounds			
Local region			
Wider region			
Distant locality			
Interstate			
Overseas			

2. What difficulties did you find in making decisions?

3. How did you resolve these difficulties?

4. What lessons can you take from this exercise when planning fieldwork in the future?

Fieldwork teaching strategies

Fieldwork teaching is inherently different from classroom teaching. The teacher is not there to simply communicate information to a passive audience; rather, as Medzini, Meishar-Tal and Sneh (2015, p. 18) describes, 'the teacher serves as a moderator who circulates among groups of students carrying out activities and experiments in the field, helps solve problems and directs the learners based on needs that arise during the activities'.

Fieldwork invites teachers to utilise a wide range of strategies when designing a fieldwork program (see Figure 9.1). They can vary from individual tasks to small, large and whole-group tasks, all taking place within the same program. The level and delivery of instructions can vary from explicit to open ended, and to individuals and groups alike. Some tasks can be quite structured (guided), while others can be self-directed. There is no correct way to design a fieldwork program. It depends on range of factors including: topic, aims, fieldwork skills to be developed, fieldwork location, health and safety considerations, age group and characteristics of the class. Foskett (1997, p. 193) notes that 'each strategy has a place in fieldwork, and a coherent fieldwork programme for a school will contain elements of each'.

According to Oost, De Vries and Van der Schee (2011, p. 311), this continuum of teaching strategies in fieldwork has 'developed from the traditional field excursion to field research based on hypothesis testing and geographical enquiry to sensory and discovery fieldwork, reflecting different perspectives on teaching and learning'.

Figure 9.1 A continuum of fieldwork teaching and learning strategies

Work increasingly student centred →
← Work increasingly teacher centred

Teacher activities	Exposition, 'talking heads'	Tight enquiry structure	Guidance on enquiry structure	Support and encouragement
Student activities	Listening, recording	Follows instructions for individual/group work	Negotiation on methods (data collection, presentation, hypothesis to test)	Students develop individual/group enquiry approach, methodology (data collection, presentation)
Fieldwork approaches	← Cook's tour approach		Data collection / hypothesis testing →	
Year level	Junior ←		Senior →	

Source: Adapted from Bartlett & Cox (1982), as described in Foskett (1997).

Short-answer questions

1.   What is sensory fieldwork and what is discovery fieldwork?

2.   How do they differ from more traditional forms of fieldwork?

3.   How or where could they have a place in your teaching and learning repertoire?

Technology

Technology in fieldwork has been around from the beginning. Geographers have always used pen or pencil and notepad to record observations and measurements, supported by successive improvements in clinometers, theodolites, compasses and cameras. Many of these 'old' technologies have been replaced or added to (e.g. GPS, GIS, laptops and now smartphones – see Chapter 6 for a more detailed discussion).

Not all technology has made its way into school fieldwork. Enterprising teachers have successfully adopted some of these technological tools into school fieldwork; however, they are not universally used. Recent developments in the use, availability and suitability of smart phones and fieldwork apps have revolutionised the fieldwork tools at the disposal of students. Chang et al. (2012, p. 44) comment that smartphones and associated applications can support student fieldwork through 'data collection, data sharing, analysing data, synthesising data and presenting results'. Furthermore, they encourage 'collaborative learning by allowing learners to communicate with one another and engaging with the environment' (Chang et al. 2012, p. 44).

Smartphones and apps can contribute to the learning that occurs during fieldwork in three ways:

1.   **Communication:** with individuals, small groups or teams, and the class through text, voice, and video.

2.   **Information creation:** photographing, documenting, and recording observations, numerous types of measurement and spatial data, and real-time sharing of this.

3.   **Information consumption:** access to the internet in real time enables students to augment their work in the field (Medzini, Meishar-Tal & Sneh 2015).

To maximise the effectiveness and use of mobile technologies, several attributes have been identified:

- Photographs need to be geo-coded.

- Quantitative data and qualitative observations must be transferable in a range of formats.

- Voice-recording capabilities must exist for specific comments of observer or interviewee.

- Real-time collaborative dataset development is necessary – both quantitative (e.g. measurements) and qualitative (e.g. surveys).

- Good quality internet access must be available.

- Apps which are easy to use and have 'little or no learning curve' (Kalyani 2012) are vital.

New digital technologies available now will facilitate more effective data collection in the field, increase the range of data that students can collect and potentially broaden the variety of possible fieldwork sites and investigations open to schools.

Practice of fieldwork

The development of a sequential and skills-based fieldwork program across all year levels is an achievable goal. Unfortunately, the development of such programs can suffer from a range of 'inhibitors' (see Table 9.5) which include:

- 'formal' inhibitors – those more legal and professional constraints that must be met as part of a teacher's responsibility in taking students out of school

- 'informal' inhibitors – those that can appear as real barriers to fieldwork, but that can be resolved appropriately – especially with experience.

Table 9.5 Inhibitors to the development of a systemised fieldwork program

Formal	Informal
Awareness of legal dimensions	Planning of fieldwork difficult; fear and concern for health and safety of students
Teacher's responsibilities	Transportation requirements
Experience and training of accompanying staff	Inexperience of teachers with fieldwork, their confidence and expertise in teaching outdoors
Pupil-to-teacher ratios	Large classes, student behaviour
Financial constraints (costs of fieldwork, staffing)	Perceived lack of resources and insufficient support
Demands of the curriculum, requirements of school curricula and timetables	Shortages of time, impact of missed lessons by supervising teachers, students missing other subjects' classes

Source: Based on ideas presented in Cranby (2002), Foskett (1997, p. 197) and Oost, De Vries & Van der Schee (2011, p. 310).

Careful, considerate and appropriate planning can allay many of the barriers that can surface. The first time a fieldwork activity is planned and carried out is the most challenging. Repeating this fieldwork activity in the following semester or year with different groups becomes simpler each time. Reflecting and learning from each successful fieldwork exercise builds experience and makes new exercises easier to plan.

The five Ps of fieldwork planning (paperwork, planning, preparation, practice, product) provide an overview of the areas teachers need to consider when planning any fieldwork activity. Taken as a whole, they may make the planning process look daunting. However, not all are relevant to every fieldwork activity.

Paperwork

This administrative stage may seem a bit tedious at first; however, once completed a couple of times, the process becomes familiar as previous activities provide a template for future ones. Remember to check the paperwork carefully each time, whether the activity is new or repeated.

School processes

Every school, education sector (government or independent) or state jurisdiction has slightly different processes in place for approval of fieldwork activities. These vary in terms of the type and amount of paperwork required, whether the activity is carried out on the school grounds, local area or further afield, and whether the activity occurs within school hours or beyond, or requires overnight stays. Although variations exist, there are some standard steps to be undertaken:

- **Initial planning.** This involves consulting the school calendar, teaching team, subject, level and school events coordinators, and the business manager.

- **Camp details.** These may include details of transport, entrance fees, accommodation and so on. A detailed budget is generally required, which may include staffing and student costs.

- **School Council approval.** Fieldwork activities taking place during the school hours or over a day are generally approved internally. Those involving one or more nights require School Council approval, generally six to eight weeks prior to the event.

- **Emergency management or departmental notification.** This may be a step in the school internal process.

- **School trip documentation.** Information for students and parents.

Often, each step requires different pieces of paperwork. Most schools have now graduated to digitally based systems, which follow similar steps to those outlined above. A word of caution, however: such generic systems, designed around the needs of everyday 'excursions', often provide insufficient details from the perspective of fieldwork. For example:

- The budget may only request input on main item summaries. This may meet the needs of the school, but not those of the teacher planning the fieldwork.

- Parent letters often provide minimal information for the parent. For legal reasons, fieldwork requires more specific information – for example, the types of activities, footwear required and supervision provided.

Early planning is paramount to schedule fieldwork programs in a school. It is important to have the planned dates, especially those involving more than one day (overnight), entered on the school calendar at least a semester or even a year ahead.

Departmental requirements

Education departments in all states fully understand their legal responsibilities with regard to camps and excursions, and cover themselves through innumerable regularly updated policies and guidelines. The responsibility falls upon the teacher-in-charge to be aware of and up-to-date with these, and adhere to them. The primary question underlying all fieldwork planning is: If one can anticipate something happening, what steps were taken beforehand to avoid or mitigate its possibility?

Each jurisdiction has its own terminology for its relevant policy, guidelines and resources. Policies may cover 'excursions', 'camps' and 'adventure activities'. Guidelines provide 'supervision and staffing ratios' and 'approval processes'. Resources can include an online notification portal or pro-forma examples of documentation (e.g. 'approval forms').

Guidelines cover all situations one might experience during a fieldwork activity. Not all are relevant to each activity, so the teacher-in-charge needs to select those appropriate to a particular activity. Resources are useful planning tools, providing suggestions or pro-formas that can be adapted to any particular fieldwork activity.

Pause and think

Explore links for your state's Camps and Excursions requirements:

1. policies

2. guidelines

3. resources.

Develop a resource folder including names and links (URLs) for the important pages.

Student information

The type and duration of a fieldwork activity determine the amount of information and permission sheets required. These can include:

- school's excursion/camp permission form – standard to school, often based on departmental requirements, generated by school's online management system when the 'event' is entered

- letter of explanation to parents – when generated by the school's online management system, can often lack the details required for the planned fieldwork activity (see following explanation)

- medical, asthma, dietary forms – standard to the school and based on departmental requirements

- clothing/equipment list – prepared by camp leader, specific to each particular fieldwork activity.

Parent letters generated by a school's online management system provide brief, minimal descriptions of the activity. They are sufficient for short excursions and outings, but lack the depth required to fully inform parents about fieldwork activities, including:

- curriculum context – where it fits in the teaching and assessment of the course (justification)

- spatial location – fieldwork can be located in more than one place and environments vary

- travel and transport arrangements – not just to the location but between and within sites visited and the form(s) of transport used, whether buses, cars, public transport or by foot

- student activities – how and in what circumstances students will be carrying out tasks, including comments on supervision (how it may vary)

- finance concerns – fieldwork is expensive, so provision for any situation (e.g. financial stress) needs to be considered and parents informed.

Not all fieldwork activities require the same amount of detail. Because of their legal responsibilities, a teacher-in-charge should ensure that parents are fully informed about and aware of what is planned.

Local authorities

When planning any fieldwork, it is advisable to contact any local authorities with responsibility over the site(s) you plan to visit. They will assist you to identify any potential access or risk potential, any restrictions on timing of visits and numbers permitted. In some cases, visitor permits may be required.

Where the fieldwork takes place in a coastal, rural or bush setting, it may be advisable to complete a trip intention form. This is necessary for any activity where a group's safety might be at risk – for example, bushwalking, places that are in remote areas, places that have extreme weather, snow or mountains, long road trips, or base camp or backpack camping trips.

A trip intention form generally contains the following details:

- name, email address and phone number of emergency contacts

- details of members of the group

- trip details, transport, start and finish locations, departure and return date and time

- contact details (e.g. mobile phone and satellite phone numbers and personal locator beacon (PLB) hex number if carried).

Each state has its own trip intention forms, prepared by police, National Parks and other groups. The Trip Intentions website (www.tripintentions.org) contains sample forms and other relevant information. Copies of trip intention plans should be left at school, in your vehicle(s) and with appropriate authorities (e.g. the police). They should all be notified upon the successful completion of your trip.

Planning

Curriculum placement

The Australian Curriculum identifies the role and place of fieldwork among the 'Inquiry and Skills' of geography from Foundation to Year 10. It stresses the collecting and processing of primary data. At the state level, some jurisdictions have refined the Australian Curriculum, placing stronger emphasis on the role of fieldwork.

Teachers planning fieldwork within their courses at any year level need to be cognisant of where it is mandated in their curricula. Furthermore, fieldwork enables teachers to meet many other skills and knowledge aims outlined in the curriculum.

Familiarity with the place of fieldwork in their curriculum enables teachers to provide:

- submissions to review and develop faculty and school-based curriculum programs

- justification for fieldwork applications to the faculty, relevant committees and responsibility positions

- parent letters clearly outlining the importance and relevance of fieldwork for their child

- responses to students, explaining the purpose and value of the fieldwork they undertake.

Secondary sources

The type, availability and number of secondary sources relevant to any fieldwork activity will vary. Secondary sources can come in a variety of formats, including:

- library reference books, map collections and other resources

- government, semi-government reports and background analysis

- websites, including site-specific (e.g. National Parks), general (e.g. local council) and action groups

- newspapers and digital media
- online digital databases (e.g. NASA), resources and applications (e.g. Google Earth)
- locally based, site-specific pamphlets and information sheets.

The use of secondary sources within a fieldwork activity will be determined by the fieldwork itself and the year level taking part. There are three opportunities to integrate secondary sources into the fieldwork learning sequence:

1. prior to the fieldwork taking place, as background and familiarisation
2. during the fieldwork, as reference (e.g. plant identification), locational (e.g. maps) and so on
3. in the write-up phase, where students access material to support their analysis of the fieldwork data obtained.

Further consideration needs to be given to the balance between teacher provision and student collection of secondary sources of data. Any use of secondary sources in a fieldwork report must be explicitly acknowledged. Reference to them in the methodology should provide a brief description and justification of their appropriateness to the fieldwork.

Pre-fieldwork reconnoitre

All fieldwork sites, be they in the school ground, local area or distant, must be visited prior to conducting the fieldwork. Tasks that need to be undertaken at this stage to ensure the orderly, safe and effective execution of the fieldwork on the day include:

- carrying out a risk assessment of all facets of the planned day
- assessing proposed vehicle routes, access to sites and availability of parking (car and bus)
- identifying areas enabling student access, walking trails or paths providing egress between sites, and availability of public conveniences
- estimating timings to enable efficient movement between sites, and sufficient time to complete activities at sites
- photographing features of sites for use in fieldwork books, in class or following fieldwork.

In preparing to repeat a fieldwork activity, the teacher-in-charge should still carry out a pre-fieldwork reconnoitre. This is necessary as conditions can change at a site. For example, access may change, there may be structural change (new buildings, old ones removed) or access to conveniences may not be available.

It is also an opportunity to meet, discuss dates and times and 'check the degree to which any activity centre or experience provider is licenced, insured, appropriately staffed' (Foskett 1997, p. 197).

Prior learning

Fieldwork can take place at the beginning, during at the end or even independently of a unit of work. Wherever it takes place, consideration needs to be given to what level of prior understanding is required for the students to successfully undertake the exercise.

The level of background knowledge will vary, depending upon where in a unit the fieldwork takes place and how the tasks are structured on the day. If the fieldwork is open-ended discovery learning, then there may be no need for focused prior learning. However, if it requires some understanding of the environment of the fieldwork, or application of specific knowledge, then prior learning is a must.

Many skills required in fieldwork tasks do not really need to be pre-taught (e.g. note-taking, observation, listening, asking questions), although students can quickly be reminded about the expectations of those skills in the context of the fieldwork. Other skills requiring more than a short explanation in the field will benefit from time in class or school grounds to develop familiarity – for example, use of GPS, conducting surveys, use of smartphone apps and completing field sketches.

Budgeting

As a general principle, no school activity can run at a loss. It is incumbent on the teacher-in-charge to ensure that each fieldwork activity finishes in the 'black'. This not only keeps the business manager happy, but careful and successful budgeting can support the endorsement of future activities.

Draw on the expertise and advice of those with experience in budgeting school activities. Do not just rely on the automated processes that are the 'norm' in your school. You need to follow appropriate policies and processes; however, you should keep a close eye on what takes place. The following are good starting points:

1. Make a detailed budget plan beforehand, accounting for all possible contingencies.

2. Once you have a total figure, calculate the per/head cost on 95 per cent of the students attending. This avoids going over budget when one or two students in a class do not attend.

3. Monitor income and expenditure as it happens, in case an adjustment is necessary. Some parts of a budget are fixed costs (e.g. transport), while others have some flexibility (e.g. food, drinks, equipment depreciation, etc.).

4. Keep detailed itemised accounts for each fieldwork activity and save budget records from year to year to help refine the accuracy of future budgets for similar activities.

Preparation

Fieldwork booklet

A fieldwork booklet is the most important element of a successful fieldwork activity – one that demands creativity, awareness, understanding, and knowledge of the background and issues of the fieldwork site. It can ensure a student's successful involvement in and completion of the fieldwork activity. A successful fieldwork booklet should empower students to address the complexities of the site(s) and meet the requirements of the curriculum.

Table 9.6 outlines the possibilities for information and resources to be included in a fieldwork booklet and suggests the extent to which such elements could be included at appropriate year levels. Ultimately, the focus, structure and development of any fieldwork booklet will be determined by the topic, site and circumstances being investigated.

Table 9.6 Fieldwork booklets – depth × year

Pages	Description	Year level					
		6/7	8	9	10	11	12
Front cover	Title, topic, location, home group, name, photograph	Simple			More complex		
Fieldwork aims	Learning outcomes and relationship to course	Simple outline of learning outcomes and assessment criteria			Clearly defined but transitional	Unit's key learnings and assessment reproduced	
Contents							
Locational map(s)	Context – school, school to site(s), site(s), campsite, etc.	As suited and requisite to the activity					
Participants	Student lists				As required for safety & records		

Table 9.6 (*cont.*)

Pages	Description	6/7	8	9	10	11	12
		Year level					
Camp rules	Rules, boundaries, etc.	Clearly delineated ←————→ increasing responsibility					
	Emergency procedures	Mandatory – appropriate to year level and activity					
Camp map(s)	Camp site, accommodation, buildings, boundaries, etc.			Varies depending on the fieldwork location, travel, and complexity			
Itinerary/ timetable		Simple			More complex		
Reference maps	Supportive material that informs the fieldwork tasks			Becoming more numerous and complex as relevant to task(s)			
Secondary data						as required	
Blank field sketches	Pro-forma, square, title bar, space to label and annotate	Half page in workflow			Full page		
Photo log	Locational and time data						
Site analysis sheets	Record sheets targeted to key knowledge objectives					Repeated analysis of different sites	
Blank maps	Used for in field recording of sites of data collection				If needed	Encouraged	
Specialised task sheets	Blank pro-forma for quadrats, transects, etc.						

Table 9.6 (*cont.*)

Pages	Description	Year level					
		6/7	8	9	10	11	12
Fieldwork sessions	Outline of tasks undertaken at identified stages	One continuous session, completed progressively		Clearly defined but transitional		Students select and carry out tasks as they proceed	
Survey	Questionnaire			Tailored to the needs and year level of the fieldwork activity			
	Recording sheets						
Lined pages	For additional note-taking			Very few	More required as complexity and duration increase		
Blank pages	To redraw maps, field sketches, etc.						

Staff briefing – expectations

Staffing fieldwork activities can be problematic, so it is worth identifying staff for the fieldwork program as early as possible. Staffing numbers are determined by the number of students, the type of environment, the activities in which they are engaged, the duration of the fieldwork and type of accommodation (if overnight).

One approach is to distribute an 'invitation to staff' when school first resumes, outlining the major fieldwork activities planned for the year. Each activity should provide information about the fieldwork theme, venue, accommodation (if overnight), dates and students (year level(s), numbers), and a brief description of what to expect.

Fieldwork activities involving a single class and of a day or less in duration will generally only require the classroom teacher themselves or occasionally one other staff member. These activities require relatively short notice, and additional staff may simply be based on availability.

When two or more staff are accompanying a fieldwork activity, day trip or overnight stay, it is important to run a briefing to outline the activity and your expectations as teacher-in-charge. This briefing can be supported by producing a staff fieldwork orientation booklet. The amount of detail required will vary depending on the complexity, duration, and student/staff numbers. The booklet's production enables the teacher-in-charge to consider and reflect upon all aspects of the planned fieldwork, including risk management and supervision.

First-aid

As teacher-in-charge, you need to ensure that both you and any accompanying staff have the appropriate first-aid qualifications and knowledge required for the fieldwork locale and activities undertaken by students. A summary of all students who have special medical requirements or needs should be provided to all staff. Fieldwork involving more adventurous activities or lacking access to quick medical assistance, may require higher qualifications. Table 9.7 outlines a range of certifications available, their renewal period and the course duration.

Table 9.7 First-aid qualifications

Certificate	Renewal	Course duration
First-Aid Level 2	Every three years	One day
Cardiopulmonary Resuscitation (CPR)	Every 12 months	Four hours
Anaphylaxis	Three years	Seek advice
Asthma Risks and Emergencies in the Workplace	Three years	
Some requirements can vary. Training may be undertaken at different or more frequent intervals. Check with your state educational jurisdiction to verify these and any possible additional requirements.		

It is important that the teacher-in-charge has the appropriate qualifications, as there may be a situation where suitably qualified staff are not available to accompany a fieldwork activity or it may involve a single class or small group. Most schools now offer the opportunity to refresh or upgrade first-aid qualifications as part staff professional development.

Fieldwork equipment

The equipment required is determined by the type of fieldwork tasks planned, and the length and duration of the fieldwork (from part- or whole-day trips to multiple day 'base camps' in more remote areas). It takes time and planning to gather and develop a supply of required fieldwork resources.

Equipment required for part- or whole-day fieldwork activities may be simple in nature (e.g. clipboards, pencils and pens, sighting poles, measuring tapes). It can be put together relatively easily through judicious use of faculty budgets, and through a small depreciation charge in the cost of the fieldwork.

'Base camps' require greater and more complex equipment that take time and resources to build up. When starting to build a program based around this type of equipment, it is best to begin with what students can bring and offer to share with the participants.

Student information in the field

On the fieldwork day, the teacher-in-charge must have with them either hard copy or electronic version of all forms (permission, medical, asthma, dietary) completed for the activity. The original forms are kept at the school. They also need to ensure that accompanying staff have copies, or at least access to the forms, if needed during the fieldwork.

Copies of student lists will also be needed by all staff taking part, for use on multiple occasions during the fieldwork (e.g. taking rolls when alighting any transport, recording group members upon completion of tasks, etc.) For some fieldwork, it may be useful for accompanying staff to have copies of student phone numbers (e.g. when working semi-independently in small groups). Similarly, students may need the number(s) of the school phone(s) being used on the fieldwork. Collection and use of student phone numbers are subject to both school and departmental protocols.

Practice

Final checklist

It is inevitable that something will slip through in the rush to pack, check people, race into admin and so on. To combat this, a simple 'loading checklist', as shown in Table 9.8, can be developed. There are always students who are willing to pack the bus, load the trailer or run a last-minute errand. However, there are often others who are left standing around not quite knowing what to do. Place the pre-prepared checklist on a clipboard, and hand it to two or three of these other students to complete. The fieldwork then begins with more of the group involved and contributing from the outset.

Table 9.8 Example fieldwork equipment: loading checklist

Item	Detail	Sighted	Loaded
First-aid kits	1 × large, 2 × small		
Esky	Ice-creams and Mars Bars		

Staff communication

During fieldwork, communication between staff is crucial for its safe and harmonious completion. The staff briefing and supporting documentation covered in school provides background for the fieldwork. Once it commences, issues arise requiring clear, up-front and ongoing communication. Examples include:

- marking of rolls – when, how often, who is responsible
- first and last off the buses, or trains – counting students on and off, checking for items left behind, etc.
- supervision of students at sites, assistance with crossing roads, movement through public spaces
- responsibility for school phone, portable first-aid kit, distribution of equipment, etc.

Field trips are dynamic environments, and issues continuously emerge, requiring quick and clear communication – most commonly as a quick standing conversation or in passing. Travelling to or between sites provides opportunities to outline what is coming up next, and to raise and discuss any issues that are likely to arise.

CONNECTION

Caterpillar method

Walking a group of students from point A to B, along a footpath and crossing roads, or in the bush along a trail, is a classic case of staff communication being essential. Such an activity should always follow the 'caterpillar' method of walking. The 'head' moves along and, after a reasonable period (given the location and topography of the walk), stops to let the 'body' catch up; it doesn't commence moving until the 'tail' appears and a head count is taken. Clear communication of who will lead, who will bring up the rear and where other staff might distribute themselves among the students is imperative.

First-aid

School first-aid kits are mostly used for sporting activities. It is incumbent on the teacher-in-charge of a geography fieldwork exercise to check that each kit contains appropriate first-aid materials. These must be appropriate to the fieldwork's location, the types of activities undertaken and the specific health or medical needs of participating students.

Two types of kits are required for geography fieldwork:

1. **Portable first-aid kit** – a smaller and more basic kit. Depending on student and staff numbers, at least one needs to be carried in a staff member's backpack during an activity.

2. **Base first-aid kit** – a more comprehensive and resourceful kit. Required for when a full class or more students are located away from immediate assistance. Usually kept on the bus next to where the fieldwork is carried out, or at the base camp when students work further away.

For extended camps working in bush environments, it is advisable to carry a lightweight fold-up stretcher. One can be made of canvas with sleeves on each long side. Two 2-metre hardwood dowel poles can be painted as sighting poles; when passed through the sleeves, they also act as carrying handles.

Record-keeping

Good record-keeping is a professional responsibility and sound practice for any fieldwork exercise. There are several things to consider.

1. **Class lists.** The 'roll call' carried out when getting on and off buses, at key meeting points and during other student movements is an essential legal record of appropriate care and management.

2. **Incident records.** If something occurs during fieldwork, a few key points must be noted immediately or at the next convenient moment. Depending on the seriousness of the incident, the amount of detail will vary.

   - Lead-up
     - Where, when and how were the expectations of the fieldwork outlined, and repeated?
     - A brief outline of relevant incidents prior to the event.
   - What occurred
     - What led to the occurrence?
     - What took place?
   - Consequences
     - What was the response of the staff at the time?
     - What follow-up steps were taken?

3. **Follow-up reports.** More serious incidents may require a formal report. Notes recorded should provide enough detail to expand into a report. Notes taken of minor incidents should be kept as future reference.

Supervision

All educational jurisdictions have strict advice surrounding student/staff ratios as they relate to different activities. The field trip will be staffed accordingly; however, the teacher-in-charge will need to be aware of changes in appropriate ratios while carrying out activities in the field. This can be achieved by applying a range of strategies in the planning stage and in the field – for example:

- clearly establishing staff expectations prior to the field trip
- designating supervision areas (yard duty)
- identifying boundaries for activities and supervision
- balancing numbers of students and staff over areas and activities.

The design of some fieldwork activities may require a degree of student independence (see examples in Table 9.9). These unsupervised activities would need the principal's approval and would only be for clearly defined instances, involve middle or senior secondary students, and occur in small groups.

Table 9.9 Graduated introduction of 'unsupervised' fieldwork

Year	Task	Management
9	City campus	Allocated a fixed precinct to explore and gather data, groups of four to five students, one mobile per group, school mobile at central base, staff moving around precincts, short time periods, then return to base for reflection.
10	Local area research	Data collection in local suburb in defined area, groups of two or three, during scheduled double period before lunch, planned tasks before leaving, sign out, sign in, show results.
11	Surveys	Door knocking in a new suburb, groups of two, allocated series of streets radiating from central point, fixed time period, school and student group mobiles, staff member at central point, other staff member circulating the area.
12	Interviews/ surveys	Rural region distant from urban school, carried out in the street, around town, staff member at central gathering point, others circulate, regular catch-up on progress, larger period of time, student free to have 'morning tea' when suits.

The teacher-in-charge would provide students with clear parameters for carrying out the activity, including maps of the sites and locations for the exercise. In the situation of independent work outside school, they would maintain records of the students involved and their time of leaving and returning. In all cases, they would ensure that students have appropriate methods of communication in the event of an emergency.

Product

All states prescribe a fieldwork report as one of their assessment tools in the senior years. They outline the knowledge and skill requirements of the fieldwork in their study designs. The assessment criteria are outlined in varying degrees of detail. They also address the use of fieldwork or primary data as a skill or technique to be developed through Years 5 to 10; however, this is not always as explicit as it is in the senior years.

Fieldwork booklet

The fieldwork booklet or resources provided to students on the fieldwork day provide an extension of the teacher's presence. The fieldwork booklet guides students in the observation and collection of appropriate data to meet the fieldwork objectives.

A fieldwork booklet's design and structure will vary with each year level, from quite structured and guided to more open-ended and self-directed. Students undertaking a school's fieldwork program from Years 6/7 to 12 would experience consistent geographic inquiry applied to increasingly demanding topics. Their skills would develop to the point where, by Year 12, they are actively involved in the design and formulation of the fieldwork.

Construction and design of questions, activities and features of a fieldwork booklet will relate to each state's curriculum, knowledge, skills and techniques as required. The depth and detail of fieldwork booklets will tend to expand from worksheets to detailed data collection, as the fieldwork expands in scale from the school, to local area, to regional and wider, over Years 7 to 12. There are, however, some general features, common to all.

In Years 6 to 8, fieldwork is mostly teacher led and structured appropriately for students' learning level. Worksheets or booklets are predominantly question or task (activity) based. Questions and activities are constructed to expose them to the geographic inquiry process, thinking and skills (e.g. observation, recording, data collection, field sketches).

By Years 9 and 10, students' fieldwork experiences move beyond school and may take place over a day or more. Fieldwork booklets become more complex and detailed. They will contain some explanation of the aims of and rationale for the fieldwork, an outline of the day's tasks, sequenced activities, questions and data-gathering pro-formas. Questions and tasks expand from relatively closed and judgemental to more open and evaluative. Booklets become self-contained (a summative assessment task) or provide data for further analysis and evaluation.

By the senior years, Years 11 and 12, fieldwork should become more self-directed, where the teacher is truly the facilitator of learning and the product provides evidence of the students' understanding and learning. In class, students familiarise themselves with the study requirements and blank data collection pro-formas designed collaboratively, ensuring that students are familiar with their purpose and use.

Fieldwork report

Data gathered in the field, including individual and shared, can now be collated, analysed and synthesised into a fieldwork report. Research skills come into play during the preparation of the fieldwork report, seeking out secondary data to support, contrast and inform the data collected in the field.

In Years 6 to 10, time for covering the curriculum is at a premium. In such cases, the fieldwork booklet becomes the assessable report. Occasionally, some collation and analysis of shared data may follow up the fieldwork itself, often referred to in the context of other class-based learning.

A post-fieldwork 'report' is more common in the senior years, although some teachers may run one in Year 10 as practice for senior level requirements. The senior curriculum's emphasis on fieldwork is often used as a template for earlier years fieldwork practices, although it will be simplified and made appropriate to the year level.

Requirements for fieldwork reports differ in depth and detail across state curricula. However, they exhibit several common elements, including:

- active engagement in the application of geographic inquiry and fieldwork skills in the collection of primary data in the field
- collation and presentation of the results of the data collected using appropriate geographic techniques
- description and analysis of the data presented
- communication of the meaning of the data (conclusions and evaluation).

Fieldwork assessment

When planning fieldwork activities, always bear in mind how you are will assess them. The manner of assessment selected will conform to the school's assessment policy. Consideration should also be given to the teaching and learning program for geography at that level.

In selecting the format of assessment, student year level, abilities and fieldwork experience all play a part. Most assessment of fieldwork takes the form of a report. Other formats could include:

- pro-forma feedback sheets – covering specific aspects of the fieldwork
- oral presentations – where individuals or small groups present to the teacher or class

- formative peer-assessment – students anonymously assessing strengths and weaknesses of peers' work (Kent, Gilbertson & Hunt 1997, p. 324).

For the early to middle years, the completion of a fieldwork booklet during the field trip can be assessed quite effectively with the use of simple, clear and targeted criteria.

Each state sets its own assessment requirements for fieldwork in Years 11 and 12, providing varying degrees of direction, from broad guidelines to quite prescriptive instructions. The trickle-down effect of the assessment process, like the structure of fieldwork described in the senior curricula tends to be adopted and mirrored to some extent at lower levels.

Short-answer questions

1. Locate your state curriculum and assessment guides for the senior years. What directions for the structure and assessment of the fieldwork component of the course do they provide?

2. How might you use this to create a fieldwork task for a Year 10 class?

3. What would the assessment criteria and scaling look like?

Fieldwork alternatives

Your fieldwork program is a major assessment item for the semester, and one or more of your students has chosen not to attend, or has been unable to do so. This is not an uncommon occurrence, so what do you do?

There are several possibilities that could lead to student non-attendance on a fieldwork activity, including:

- parental disapproval – can vary from personal views on educational efficacy of activity to cultural attitudes (e.g. if it involves overnight accommodation)

- student malady – from an illness to more personal circumstances

- timing – where the date(s) of the fieldwork clash with family or cultural events.

How does one cover for the missed opportunity? The response chosen will depend on the type of fieldwork, your expectations and the individual student. There are numerous possible responses, including:

- creating a substitute fieldwork experience (may require parental supervision) that meets both curriculum objectives and personal requirements

- alternative time where, with family support, the student undertakes an abridged version of the fieldwork, but enough that they can share and benefit from the class's collected data

- providing an alternative 'virtual fieldwork' experience, although the extent to which it replicates the fieldwork skills of the task may be problematic.

Conclusion

Fieldwork as described in this chapter is, and should be, central to every student's geography education. As geography teachers, our role within the faculty, school and wider community is to 'articulate a strong case for the presence and enhancement of fieldwork and direct experience as an integral element of all good geographical education' (Binns 1996, p. 52).

As fieldwork is clearly identified in the Australian Curriculum: Geography, and state curriculum authorities have embedded the skills and practice of fieldwork in their senior geography curricula and emphasise fieldwork as an assessment task in these curricula, geography teachers must embed the practice of fieldwork in the early secondary years to develop students' skills and experiences. One way to support this practice is to enshrine fieldwork as an integral part of the school's geography curriculum through faculty and school policy documents.

Bringing it together

1. What does fieldwork bring to geography that makes it so central to the discipline?

2. Describe three ways in which fieldwork can broaden your teaching pedagogy.

3. Briefly outline two of the more challenging aspects of incorporating fieldwork into your teaching and explain how you might mitigate that challenge.

4. Investigate a smartphone app that could be used for fieldwork. How could it contribute to the fieldwork process and students' learning?

5. What are the significant features of a 'risk assessment' that you need to consider when taking a Year 9 class on a coastal fieldwork exercise?

6. In the school in which you are based, plan a fieldwork activity for a class you teach.

   a. Identify and locate the processes, people, and paperwork with which you would need to engage.

   b. Select the fieldwork's curriculum context (area of study, content descriptor, skills, etc.).

   c. Choose an appropriate fieldwork site. Prepare a planning summary of all the external agents you will need to contact, and be aware of – for example, transport, council, local experts.

References

Bartlett, L. & Cox, B. (1982) *Learning to teach geography: Practical workshops in geographical education*, Brisbane: Wiley.

Binns, T. (1996) School geography: The key questions for discussion, in E. Rawling & R. Daugherty, eds, *Geography into the 21st century*, Chichester: Wiley.

Boardman, D. (1983) *Graphicacy and geography teaching*, Beckenham: Croom Helm.

Chang, C.H., Chatterhea, K., Goh, D.H.-L. ... Nguyen, Q.M. (2012) Lessons from learner experiences in a field-based inquiry in geography using mobile devices, *International Research in Geographical and Environmental Education*, 21(1), 41–58.

Cranby, S. (1998) A Framework for Negotiating in the Geography Classroom, *Interaction*, 16, 23–33.

——(2002) Fieldwork – a whole school approach, *Interaction*, 30, 32–8.

——(n.d) Teacher feedback during fieldwork workshops offered at AGTA, GTAV and other state GTA Conferences covering the period 1996–2002, unpublished.

Foskett, N. (1997) Teaching and learning through fieldwork, in D. Tilbury & M. Williams, eds, *Teaching and learning geography*, London: Routledge.

Haggett, P. (1995) *The geographer's art*, Oxford: Blackwell.

Huckle, J., ed. (1983) *Geographical education: Reflection and action*, Oxford: Oxford University Press.

Kalyani, C. (2012) Use of mobile devices for spatially cognizant and collaborative fieldwork in geography, *Review of International Geographical Education Online*, 2(3), 303–25.

Kent, M., Gilbertson, D.D. & Hunt, C.O. (1997) Fieldwork in geography teaching: A critical review of the literature and approaches, *Journal of Geography in Higher Education*, 21(3), 313–32.

Laws, K. (1984) Learning geography through fieldwork, in J. Fien, R. Gerber & P. Wilson, eds, *The geography teacher's guide to the classroom*, Melbourne: Macmillan, pp. 104–17.

Medzini, A., Meishar-Tal, H. & Sneh, Y. (2015) Use of mobile technologies as support tools for geography field trips, *International Research in Geographical and Environmental Education*, 24(1), 13–23.

Oost, K., De Vries, B. & Van der Schee, J. (2011) Enquiry-driven fieldwork as a rich and powerful teaching strategy – school practices in secondary geography education in the Netherlands, *International Research in Geographical and Environmental Education*, 20(4), 309–25.

Tuan, Y.-F. (1974) *Topophilia: A study of environmental perception, attitudes, and values*, New York: Columbia University Press.

Westhorpe, C. (2011) Being a geography teacher, *Interaction*, 39, 13–15.

Winchester, S. (2001) *The map that changed the world*, Harmondsworth: Penguin.

The general capabilities' synergy with geography

Malcolm McInerney

Learning objectives

By the end of this chapter, you will be able to:

- identify the place and purpose of the general capabilities in the Australian Curriculum

- understand the nature of the general capabilities of the Australian Curriculum

- illustrate the relevance and synergy of each of the general capabilities to the aims, content, understandings and skills of the Australian Curriculum: Geography

Introduction

The Humanities and Social Sciences (HASS) learning area is one of the eight learning areas of the Australian Curriculum. It includes the subjects of history, geography, civics and citizenship, and economics and business. The Australian Curriculum promotes the idea that the curriculum needs to address not only the intellectual growth of young people, but also their social, emotional, ethical, humanitarian and aesthetic development and wellbeing. When designing the Australian Curriculum, some of these goals were seen as not conveniently fitting into the traditional content of the learning areas, so it was decided to distribute them across, and immerse them in, the learning areas to ensure that they were part of the teaching and learning requirements and student entitlement of the Australian Curriculum.

Guided by the 2008 Melbourne Declaration (MCEETYA 2008), the developers of the Australian Curriculum considered that schooling needed to have 'a broader frame' with regard to educational goals for young Australians and the following needed to be addressed when designing learning in Australian schools:

- a 'respect for cultural and religious diversity' and 'the sense of global citizenship' (ACARA 2010, p. 4)

- literacy and numeracy skills

- the promotion of an active and informed citizenship with moral and ethical integrity

- problem-solving approaches in a creative ways

- information, communication and technology (ICT) skills and awareness

- the social interaction capacity of individuals and awareness of personal values and attributes

- development of confident and creative individuals

- outcomes for Aboriginal and Torres Strait Islander peoples

- an 'ability to relate to and communicate across cultures, especially the cultures and countries of Asia' (ACARA 2010, p. 7).

The outcome of such thinking was that, in addition to the learning areas, there was seen to be a need for the development of general capabilities and cross-curriculum priorities that students would be taught and learn when engaging with the Australian Curriculum. The resulting curriculum, with learning areas, general capabilities and cross-curriculum priorities, has been referred to as a multi-dimensional curriculum that aims to meet the broader curriculum frame in terms of intent and aims of the goals of Australian schooling as delineated in the Melbourne Declaration (Figure 10.1).

In December 2019, the document that was developed following a review of the Melbourne Declaration was released. The new national declaration on education goals for all Australians is known as the Alice Springs (*Mparntwe*) Education Declaration. It sets out the national vision for education and the commitment of Australian governments to improving educational outcomes through the 2020s. As was the case with the 2008 Melbourne Declaration on Educational Goals for Young Australians, the Alice Springs *(Mparntwe)* Education Declaration also has the goals of focusing on developing excellence and equity to build the capacity of citizens to be 'confident and creative individuals, successful learners, and active and informed community members' (DESE 2020).

This chapter addresses the place and nature of the general capabilities of the Australian Curriculum and emphasises that the general capabilities are not an 'add-on' to the teaching of the Australian Curriculum: Geography. Rather, they are integral to

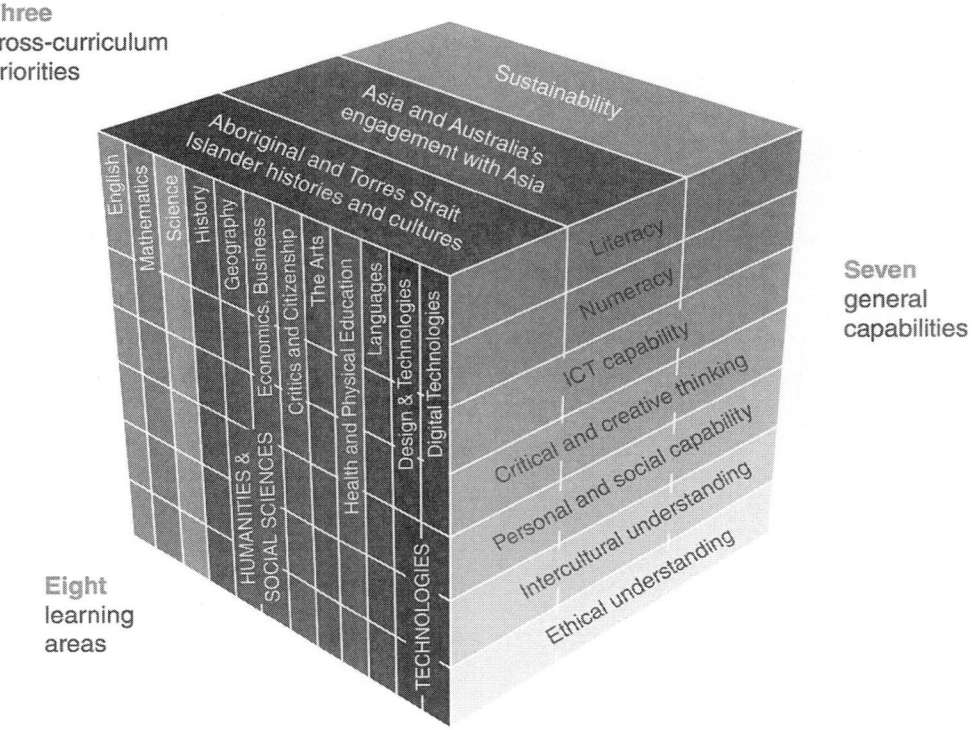

Figure 10.1 The Foundation to Year 10 Australian Curriculum is described as a three-dimensional curriculum that recognises the central importance of disciplinary knowledge, skills and understanding, general capabilities and cross-curriculum priorities

Source: ACARA (2020a).

quality geographical education and have a very comfortable synergy with the aims, knowledge, understandings, skills and pedagogical approaches of the geography curriculum.

The general capabilities

The designers of the Australian Curriculum identified seven general capabilities: literacy; numeracy; information and communication technology (ICT) capability; critical and creative thinking; personal and social capability; intercultural understanding; and ethical understanding (Figure 10.2). In this section, the general capabilities will be described in general terms, before being applied specifically to the geography curriculum.

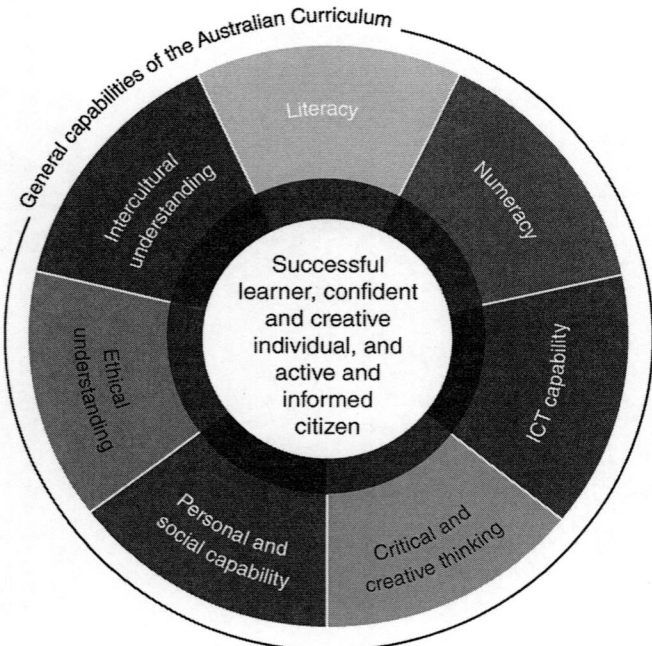

Figure 10.2 The seven general capabilities of the Australian Curriculum

Source: ACARA (2020b).

CONNECTION

The general capabilities of the Australian Curriculum

GC 1: Literacy

Students become literate by developing the knowledge, skills and dispositions to interpret and use language to effectively participate in society. Literacy efficacy involves the confidence and ability to listen, 'read, view, speak, write and create oral, print, visual and digital texts, and use and modify language for different purposes in a range of contexts' (ACARA 2020b).

GC 2: Numeracy

Students become numerate by developing the knowledge and skills to use mathematics confidently in their studies and in everyday life. Numeracy efficacy aims at students having the capacity to use numeracy in a wide range of situations, understanding the role of numeracy in the world and having the capacity to 'use mathematical knowledge and skills effectively' (ACARA 2020b).

GC 3: Information and communication technology (ICT) capability

Students develop the ICT capability by learning to use ICT effectively and appropriately to access, create and communicate 'information and ideas, solve problems and work collaboratively in all learning areas at school and in their lives beyond school' (ACARA 2020b).

GC 4: Critical and creative thinking

Students develop the two strongly linked skills of 'critical and creative thinking as they learn to create and evaluate knowledge, clarify concepts and ideas, investigate possibilities, consider alternatives and solve problems' (ACARA 2020b) (Figure 10.3). They are encouraged to think widely and in depth by employing skills such as reasoning, 'logic, resourcefulness, imagination and innovation' in their studies and everyday life' (ACARA 2020b).

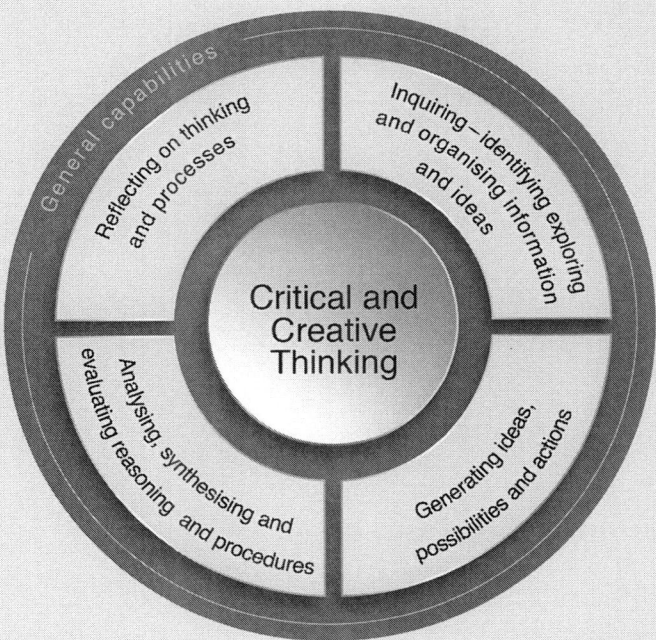

Figure 10.3 Diagrammatic representation of the critical and creative thinking general capability

Source: ACARA (2020b).

GC 5: Personal and social capability

Aims for students to 'understand themselves and others, manage their relationships, lives and work and to learn more effectively' (ACARA 2020b). This involves students recognising 'emotions, developing empathy for others and understanding relationships, making responsible decisions, working effectively in teams, dealing with challenging situations constructively and developing leadership skills' (ACARA 2020b) (see Figure 10.4).

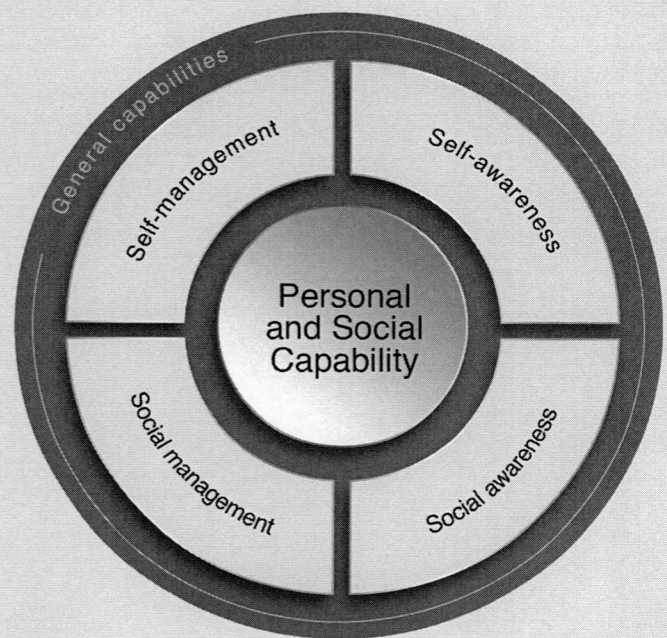

Figure 10.4 Diagrammatic representation of the personal and social general capability

Source: ACARA (2020b).

GC 6: Intercultural understanding

Students 'learn to value their own cultures, languages and beliefs, and those of others' (ACARA 2020b). The capability focuses on students understanding 'how personal, group and national identities are shaped, and the variable and changing nature of culture' (ACARA 2020b) (see Figure 10.5).

GC 7: Ethical understanding

Students 'identify and investigate the nature of ethical concepts, values and character traits' and understand how 'reasoning can assist ethical judgement and awareness of the influence that their values and behaviour have on others' (ACARA 2020b) (see Figure 10.6).

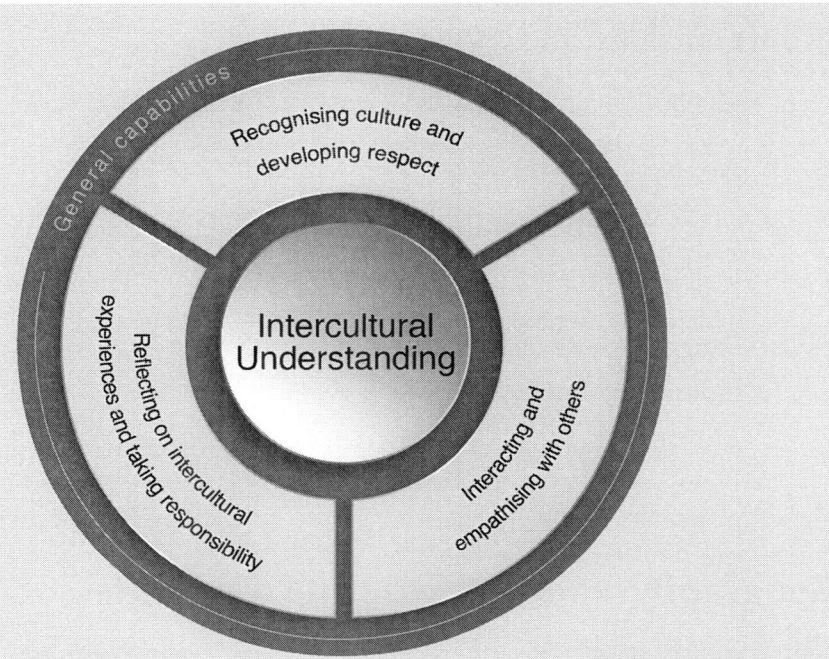

Figure 10.5 Diagrammatic representation of the intercultural understanding general capability

Source: ACARA (2020b).

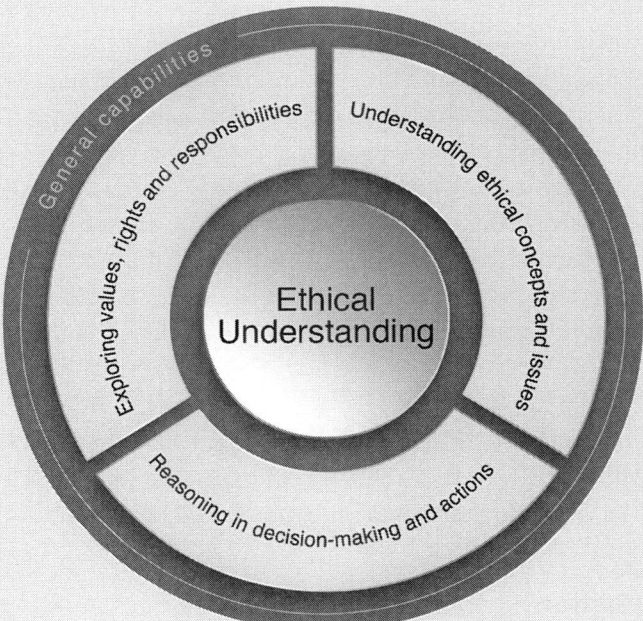

Figure 10.6 Diagrammatic representation of the ethical understanding general capability

Source: ACARA (2020b).

Short-answer questions

1. Identify a problem in your local area that will require students to engage in a classroom inquiry project to research, think critically about and develop a creative solution to resolve/solve.

2. What is the difference between a conversation and discussion? Which of the two do you think is most important in the geography classroom and why?

3. Think of ways in which you could redesign your classroom and teaching processes to encourage greater interaction between students and improve opportunities for discussion and conversations.

4. Think of an issue aligned to the geography curriculum at a particular year level that could be used in the classroom to develop students' ethical thinking.

Geography and the general capabilities

You would have noted that in the summaries of the general capabilities above, detailed diagrams were not displayed for the general capabilities of literacy, numeracy and ICT (GC 1–3) but were displayed for critical and creative thinking, social and personal capability, intercultural understanding and ethical understanding (GC 4–7). This was in recognition of the fact that GC 1–3 apply to geography as well as all other learning areas in a variety of legitimate and comprehensive ways. However, it is considered that GC 4–7 have a very special place in geographical education and require greater consideration as critical capabilities that should be developed to a high degree in the geography classroom. That is not to say that other learning areas do not address these capabilities; however, it can be argued that the general capabilities of critical and creative thinking, social and personal capability, intercultural understanding and ethical understanding have a significant synergy with the aims, content and skills developed in geography, as they do for all the HASS subjects.

As indicated by the presence of icons throughout the online presentation of the Australian Curriculum (Figure 10.7), the general capabilities are addressed through the content descriptions of the learning areas.

The general capabilities in the Australian Curriculum learning areas are identified 'wherever they are developed or applied in the content descriptions' (ACARA 2020a). They are also identified where they offer 'opportunities to add depth and richness to student learning in the content elaborations' (ACARA 2020a).

Literacy in geography

The Australian Curriculum: Geography is a language-rich subject. As a result, students are required to engage with written, graphed and drawn materials. Texts may include

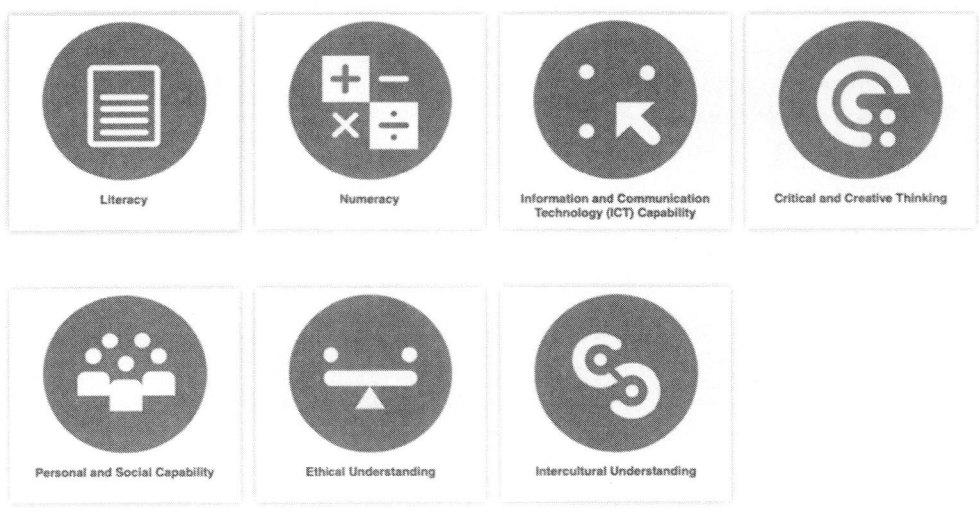

Figure 10.7 The General Capabilities icons used in the online curriculum

Source: ACARA (2020b).

'narrative accounts, reports, explanations, arguments, debates, maps, tables, graphs and images' (ACARA 2021). These sources of information include geography-specific vocabulary for students to 'make meaning' of and evaluate texts to formulate an opinion. Supporting the development of literacy in the geography classroom will involve students in debates and discussions to identify and examine points of view when developing conclusions (ACARA 2021).

The importance of literacy skills is particularly evident in the Inquiry and Skills sub-strand of the geography curriculum (ACARA 2020b). For example, at Year 9 as part of the inquiry process, students are to be involved in the literacy-based challenges of:

- developing significant geographical questions (ACHCS063)

- evaluating sources for their reliability, bias and usefulness (ACHCS064)

- presenting findings, arguments and explanations in a range of appropriate communication forms, selected for their effectiveness and to suit audience and purpose; using relevant geographical terminology, and digital technologies as appropriate (ACHCS070).

Numeracy in geography

In geography, students develop the 'numeracy capability as they apply numeracy skills in relation to geographical inquiries' (ACARA 2020b). Geography is particularly rich in numeracy through data use, mapping and fieldwork, requiring data collection, the construction and interpretation of graphs, 'maps, models, diagrams and satellite images, working with the space concepts of grids, scale, distance, area and projections', locational specification and analysis, and distance measurements (ACARA 2021). These numeracy skills are considered important to develop graphicacy skills in geography.

The importance of the numeracy skills is particularly evident in the Inquiry and Skills strand of the geography curriculum (ACARA 2020b). For example, at Year 9 as part of the inquiry process, students are to be involved in the numeracy-based challenges of:

- representing multi-variable data in a range of appropriate forms, for example scatter plots, tables etc (ACHCS065)

- interpreting and analysing multivariable data and other geographical information using quantitative methods, to make generalisations and inferences, propose explanations for patterns, trends, relationships and anomalies, and predict outcomes (ACHCS067)

- presenting findings, arguments and explanations in a range of appropriate communication forms (ACHCS070).

ICT capability in geography

In geography, 'ICT capability involves the student capacity to locate, process, analyse, evaluate and communicate data and information using digital technologies' (ACARA 2021). Such technologies include geographic information systems (GIS) and remote-sensed visualisations such as Google Earth. ICT is also important in the geography classroom for students to locate, represent and analyse information to develop conclusions and possible futures. Students need to develop an awareness of the positive and negative impacts of technologies on places and human activity.

The importance of ICT is particularly evident in the Inquiry and Skills strand of the geography curriculum (ACARA 2020b). For example, at Year 10 as part of the inquiry process, students are to be involved in the use of ICT by:

- representing multivariable data in a range of appropriate forms, with the use of digital and spatial technologies (ACHCS074)

- representing spatial distribution of geographical phenomena by constructing special purpose maps using spatial technologies as appropriate (ACHCS075)

- interpreting and analysing multivariable data and other geographical information using qualitative and quantitative methods, and digital and spatial technologies as appropriate (ACHCS076)

- identifying how geographical information systems (GIS) might be used to analyse geographical data and make predictions (ACHCS078)

- presenting findings, arguments and explanations using relevant geographical terminology and digital technologies as appropriate (ACHCS079).

Critical and creative thinking in geography

An important and overt aim of geography is for students to develop critical and creative thinking as they inquire using the geographical concepts. As investigators, inquirers, problem-solvers and conversationalists, students in geography are encouraged to develop a conceptualised approach to their learning, and in the process become critical and creative thinkers. Such thinking involves the framing of 'rich' inquiry questions, critiquing and analysing sources, 'using evidence to develop arguments, interpreting and analysing data, and systems thinking' to make recommendations and consider possible, probable and preferred futures (ACARA 2021). The geography student thinks conceptually and deeply about questions posed and develops the realisation that no question is simple and a multiplicity of points of view and possible solutions usually exist in relation to any one issue. Such an approach in geography requires students to be inquisitive and creative as they analyse, speculate and develop interpretations to explain features, events and phenomena in the past, present and future. For example, critical and creative thinking can be applied in geography to find solutions to traffic congestion in the CBD of a large city.

The importance of critical and creative thinking is particularly evident in the Inquiry and Skills strand of the geography curriculum (ACARA 2020b). For example, at Year 10 as part of the inquiry process, students are to:

- develop geographically significant questions and plan an inquiry (ACHCS072)

- evaluate sources for their reliability, bias and usefulness (ACHCS073)

- apply geographical concepts to synthesise information from various sources and draw conclusions based on the analysis of data and information, taking into account alternative points of view (ACHCS077)

- reflect on and evaluate findings of an inquiry to propose individual and collective action in response to a contemporary geographical challenge; and explain the predicted outcomes and consequences of their proposal (ACHCS080).

Personal and social capability in HASS

Geography is a learning area that focuses on the question of 'being human' on this planet. Hence, understanding about people, places, processes and phenomena is a capability that geography educators are very comfortable about embracing as an important aspect of learning. Implicit in the geography classroom are the practices of collaboration, reflection, communication, 'appreciation of the insights and perspectives of others and the exploration of personal identity' and sense of belonging (ACARA 2021).

Affirming the goal of developing active and informed citizens, the 'reflecting and responding' inquiry stage in the geography curriculum aims at providing the opportunity for students to 'make a contribution to their communities and society more broadly'

(ACARA 2021). Such capacity-building involves the development of 'leadership, resilience, goal setting, advocacy skills and informed and responsible decision-making by individuals' (ACARA 2021). For example, the personal and social capability can be applied when students are working in teams conducting fieldwork on excursions in geography.

The importance of personal and social capability is particular evident in both the Knowledge and Understanding, and Inquiry and Skills strands of the geography curriculum. For example, in the Year 8 Knowledge and Understanding strand, students learn about:

- spiritual, aesthetic and cultural value of landscapes and landforms for people (ACHGK049)

- management and planning of Australia's urban future (ACHGK059).

In the Inquiry and Skills strand, students are challenged to:

- present findings, arguments and ideas in a range of communication forms selected to suit a particular audience and purpose (ACHCS061)

- reflect on their learning to propose individual and collective action in response to a contemporary geographical challenge (ACHCS062)

- explain the predicted outcomes of a proposal (ACHCS062).

Intercultural understanding in HASS

In geography, students develop intercultural understanding by learning about the diversity of the world's places and the people who occupy those places. By studying their own and others' cultures and world-views, students gain an understanding of the complexity of cultures and the interaction and interdependencies between cultures both within and between countries. Through the study of the significance of place to cultural identity, students gain an understanding of cultural values and experiences for groups and learn about the nature of interaction across cultural boundaries. Such studies involve the exploration of how being a member of a group, values and customs impact a society.

For example, in the Knowledge and Understanding strand of Year 10 geography (ACARA 2020c), students learn about:

- the approaches taken by Aboriginal and Torres Strait Islander peoples to custodial responsibility and environmental management in different regions of Australia (ACHGK072)

- different ways of measuring and mapping human wellbeing and development, and how these can be applied to measure differences between places (ACHGK076)

- reasons for, and consequences of, spatial variations in human wellbeing on a regional scale (ACHGK079)

- issues affecting development of places and their impact on human wellbeing (ACHGK078).

CONNECTION

A case study to develop student cultural understanding for the Year 8 geography Landforms and Landscape unit

Content description: Spiritual, aesthetic and cultural value of landscapes and landforms for people, including Aboriginal and Torres Strait Islander peoples (ACHGK049)

This case study enables students to deconstruct and analyse cultural perception and understanding of the issue surrounding the climbing of Uluru by tourists to Central Australia. For many years, the topic of climbing Uluru was an issue, with stakeholders on both sides arguing their case. The tourist industry saw the climbing of Uluru as an important drawcard for attracting tourists from around the world to Central Australia, while the Aboriginal people of the area saw it as a culturally inappropriate treatment of the sacred landform. On 26 October 2019, 35 years after the local Anangu people were granted land rights to the area, the climbing trail was finally closed. As a case study, this is an excellent opportunity for students to discuss the cultural nuances of the issue and gain an understanding of the view of the Aboriginal people regarding the cultural significance of landmarks such as Uluru.

Pause and think

Choose one of the capabilities listed below. For one of the proposed content descriptions listed for that capability, develop an elaboration that does not appear in the curriculum, which could be taught in the geography classroom to facilitate the learning for that capability:

1. Critical and creative thinking

2. Personal and social

3. Intercultural understanding.

For example, critical and creative thinking could be developed when studying the Geographies of Interconnections Unit in Year 9, focusing on the *effects of people's travel, recreational, cultural or leisure choices on places, and the implications for the future of these place* content description. This content description provides the opportunity for students to think creatively about a future impacted on by changing travel, lifestyle choices and cultural changes in the twenty-first century. Such thinking should also encourage students to be critical as they analyse and evaluate the positives and negatives of possible, probable and preferred changes in the physical and human environments into the future.

An excellent elaboration for this topic would be the impact of COVID-19 on the movement of people around the world. Students could look at the impact of travel restrictions on cultural sharing, communications between countries as a result of diminished global interconnections and examine impacts, discussing whether they are positive or negative for local, regional and global futures.

Ethical understanding in HASS

Through geographical inquiries, students develop an understanding of the ways in which diverse values and principles influence human activity and the views of individuals and groups. Through the embedding of this capability in geography, students develop informed, ethical values and attitudes through the study of social, economic, political and 'environmental issues. This will involve students examining their roles, rights and responsibilities' as informed citizens (ACARA 2021). For example, the ethical understanding could be applied in geography when considering the ethics of developed countries advocating that less-developed countries should limit car ownership due to ever-increasing global pollution levels.

For example, in the Year 10 geography Knowledge and Understanding strand (ACARA 2020c), students learn about:

- environmental world-views of people and their implications for environmental management (ACHGK071)
- issues affecting development of places and their impact on human wellbeing (ACHGK078).

In the Inquiry and Skills strand, students are to:

- evaluate sources using ethical protocols (ACHCS073)
- reflect on and evaluate findings of an inquiry to propose individual and collective action in response to a contemporary geographical challenge, taking account of environmental, economic, political and social considerations (ACHCS080)
- explain the predicted outcomes and consequences of a proposal (ACHCS080).

CONNECTION

A case study to develop student ethical thinking for the Year 8 geography Changing Nations unit

Content description: Reasons for, and effects of, international migration in Australia (ACHGK058)

This case study enables students to explore the values and ethics behind the issue of international immigration to Australia. As a case study, it needs to go beyond studying the number of migrants, types of migrants and destination of migrants in Australia. The case study will need to explore the processes involved in the treatment of new arrivals and issues associated with the types of migrants (detention, asylum seeking, etc.) and settlement of migrants into the Australian community. Such an exploration leads to ethical discussions associated with the source of migrants, ease of entry, settlement options and refugee numbers. Furthermore, students will inevitably have the opportunity to discuss ethical issues related to the diversity of Australian society as a result of migration (source of migrants) and the size of the Australian population (big/small Australia debate) in relation to migration numbers.

CONNECTION

A case study to develop student ethical thinking for the Year 10 geography Geographies of Human Wellbeing unit

Content description: Reasons for spatial variations between countries in selected indicators of human wellbeing (ACHGK077)

This case study enables students to examine the ethical aspects of governments initiating population controls to reduce population growth in the region. Overpopulation and the continual growth of population in less-developed countries intensifies issues of poverty and resource use for such countries. Over the years, governments in these countries have attempted to reduce population growth through a range of measures that raise ethical issues regarding individual freedoms and human rights to have children and the sustainability needs of the country. China's former one child policy is an excellent example of such an ethical dilemma for students to examine the personal needs of a people and the environmental and economic needs of a country. Other ethical dilemmas associated with this topic involve the examination of the sterilisation policies in countries such as Bangladesh in the 1980s and the use of contraception in less-developed Catholic-dominated nations.

Short-answer questions

1. When does a topic to be studied became an issue in geography? What are the characteristics of a topic that make it an issue?

2. Suggest a range of issues that could be discussed in the geography classroom to develop students' ethical understanding.

3. Are all issues in geography actually problems to be solved?

4. Does a geographical topic need to be an issue to develop ethical thinking and intercultural understanding?

5. Is there a danger of creating a problem-based and negative teaching program in geography if too much emphasis is placed on identifying and analysing issues? Why or why not?

Conclusion

Geography educators need to be comfortable with the idea that all the general capabilities are soundly and validly embedded within their learning area. There is considerable synergy with the aims of geographical learning and the intent and capacity building implicit in the general capabilities. While all the general capabilities are of relevance to teaching and learning in geography, those of critical and creative thinking, personal and social capability, intercultural understanding and ethical understanding are highly visible components of quality geographical education. One of the primary goals of the Melbourne Declaration was for schooling to develop informed and responsible citizens, a goal supported by the development of student capabilities with which geography educators strongly identify.

This chapter has described the nature of the general capabilities and articulated their place in the geography curriculum and classroom. All the general capabilities are visible in geography in a meaningful and extensive way and in no way do geography educators have to 'fit a square peg in a round hole' to address the general capabilities or create tokenistic connections to the curriculum to justify a discussion of the place of the general capabilities in geography. There is a strong synergy between the aims of the geographical learning and the intent of the general capabilities to develop young people who have the capabilities of being informed, responsible and active social, economic and environmental citizens. Geography has always played, and will continue to play, an important role in a balanced school curriculum to deliver these important environmental, social, economic and political capabilities in a meaningful and contextually relevant manner, so young people can develop to their maximum capacity.

Bringing it together

1. Why is the Australian Curriculum referred to as a three-dimensional curriculum?

2. What are the general capabilities of the Australian Curriculum?

3. Choose two of the general capabilities and discuss how they can be said to reflect some of the core purposes of geographical learning.

4. For a particular geography year level, identify content descriptions of the Knowledge and Understanding or Inquiry and Skills strands that can be seen as relating to the general capabilities in a meaningful way. Explain the reasons for your choices.

5. There is considerable debate across Australia regarding the question of whether the general capabilities should be overtly taught and possibly even part of the formal assessment of the Australian Curriculum. What is your view on this question and how do you think they could comprise part of the assessment of student performance when studying the Australian Curriculum: Geography?

References

Australian Curriculum, Assessment and Reporting Authority [ACARA] (2010) *The Shape of the Australian Curriculum*, Version 4.0, Canberra: ACARA, retrieved from https://docs.acara.edu.au/resources/The_Shape_of_the_Australian_Curriculum_v4.pdf

——(2020a) *F–10 Curriculum – Structure*, Canberra: ACARA, retrieved from www.australiancurriculum.edu.au/f-10-curriculum/structure

——(2020b) *F–10 General capabilities*, Canberra: ACARA, retrieved from www.australiancurriculum.edu.au/f-10-curriculum/general-capabilities/

——(2020c) *F–10 Curriculum: Humanities and Social Sciences, Geography*, Canberra: ACARA, retrieved from www.australiancurriculum.edu.au/f-10-curriculum/humanities-and-social-sciences/geography

——(2021) *General Capabilities in the Australian Curriculum: Humanities and Social Sciences*, Canberra: ACARA, retrieved from https://docs.acara.edu.au/resources/HASS_-_GC_learning_area.pdf

Department of Education, Skills and Employment [DESE] (2020) *The Alice Springs (Mparntwe) Education Declaration*, retrieved from https://docs.education.gov.au/node/53193

Ministerial Council on Education, Employment, Training and Youth Affairs [MCEETYA] (2008) *Melbourne Declaration on Education Goals for Young Australians*, retrieved from www.curriculum.edu.au/verve/_resources/National_Declaration_on_the_ Educational_Goals_for_Young_Australians.pdf

The importance of planning in geography

Malcolm McInerney, Susanne Jones and Susan Caldis

Learning objectives

By the end of this chapter, you will be able to:

- develop a planning approach in geography by identifying the learning requirements for the chosen unit of the Australian Curriculum: Geography

- explore the nature of quality assessment in geography, including the collection of student evidence of geographical learning

- consider a way to plan a unit of work in geography

Introduction

This chapter aims to develop teacher skills in planning units of work in geography. To do this, teachers must become aware of the learning requirements of the unit from the Australian Curriculum: Geography that they intend to teach and then be focused on the assessment aspects that will drive the planning for the unit. Finally, the chapter provides a planning approach for the development of a teaching program that is relevant, achievable and engaging for teaching.

The planning approach

The chapter draws from the work of Grant Wiggins and Jay McTighe (2005) in believing that teachers are designers and that a key aspect of a teacher's work is to design curriculum, learning experiences and assessment that will meet clearly identified learning requirements and needs of students – that is, for quality planning to

> guide our teaching and to enable us, our students, and others (parents and administrators) to determine whether our goals have been achieved, that is,

did the students learn and understand the desired knowledge? (Wiggins & McTighe 2005, p. 1)

Considered and strategic planning is critical for quality and relevant learning. While true for all subjects taught, it is especially important in geography because of the breadth of content in this holistic subject. Without targeted planning, the identification and teaching of geographical content can be very daunting tasks, so it is important that the teacher is very clear about the purpose, knowledge and skills to be focused on in a unit of work. The other critical stage of the process of planning is that when the teacher is clear about the required knowledge, conceptual thinking, skills and overall purpose of the unit, they must be at front and centre before activities, resources, assessment criteria and assessment tasks are determined. We can refer to such an approach as front-end planning, based on backward design. The understanding by design (UBD) work of Wiggins and McTighe (2005) has been influential in curriculum planning and pedagogical approaches in recent years, and forms the basis of much of the thinking adopted and approach taken in this chapter. This chapter is not the place to go into great detail about UBD, but it is the methodological approach at the core of the processes that will be described.

CONNECTION

The nature of backward design

Some key thinking from the writings of Wiggins and McTighe (2005) relevant to design and planning units of work is outlined below.

Teachers are designers

- The three stages of backward design are:

  1. identify desired result

  2. determine acceptable evidence

  3. plan learning experience and instruction.

- Good design is about learning to be more thoughtful and specific about our purposes and what they imply.

- The shift involves thinking a great deal about the specific learnings sought and the evidence of such learning, before thinking about what we, as teachers, will do or provide in teaching and learning activities.

- The challenge is to focus first on the desired learnings from which appropriate teaching will logically follow.

- Best designs derive backward from the learnings sought.

- Too many teachers focus on the teaching and not the learning.

- Results-focused design challenges the traditional content-focused design.

- Focus on the 'Why?' and 'So what?' questions as the focus of curriculum planning.

- The twin sins of activity-oriented design (hands-on without being minds-on) and aimless coverage need to be avoided.

- Students require clear purposes and explicit performance goals.

To model UBD in relation to the planning of geographical learning, this chapter is organised to demonstrate the following:

- deconstructing the curriculum content descriptions to identify the learning requirements (desired result)

- determining quality assessment and identifying the assessment aspects for the unit (acceptable evidence of learning)

- designing the unit with the requirements and assessment aspects front-ended (planning learning and teaching approaches).

Preparing for planning: Conceptualising the thinking

What do we really want our students to learn? When looking at the written curriculum, teachers as curriculum-makers need to intellectualise the essence of the topic to identify what really is at the core of the learning required. It is not about coverage and ticking boxes, but rather developing an overall view of the curriculum and to develop a holistic approach in regards to knowledge, understandings, skills and conceptualisation. Lambert (2016a) refers to such work by teachers as the 'creative act of interpreting a curriculum and turning it into a coherent, challenging, engaging and enjoyable scheme of work, it is a job that really never ends and lies at the heart of good teaching'. A pivotal aspect of curriculum-making is that it recognises the importance of the needs, motivations and experiential knowledge of students. Considered planning provides a clear understanding of the curriculum and possible pedagogical approaches, and demonstrates the distinction between the children's experiences and disciplinary knowledge.

Lambert (2015) provides a useful model that can inform unit planning by teachers and help to reconcile and connect aims-led curriculum planning and knowledge-led curriculum planning. As Michael Young (2013) highlights in his work on the powerful knowledge of geography, there is knowledge (knowledge, understanding and skills) in geography that is not necessarily just absorbed by individuals in everyday life, but rather needs to be overtly taught to ensure that all citizens receive the benefits of such knowledge.

When curriculum-making from the written curriculum, teachers need to be cognisant of the powerful knowledge implicit in the curriculum, teaching strategies available to engage students in learning and the need to recognise where the student is at and what their needs are. This is not an easy task, but it is certainly critical if geographical planning is to be relevant, meaningful and achievable for students – so the teacher is not just a 'boffin' keen to share their knowledge and passions (Lambert 2015).

The curriculum making model developed by Lambert (2016b) provides a visualisation for planning, where the aim for the teacher is to develop a balance between the competing priorities of the students, the teaching and the subject (Figure 11.1). Good curriculum-making occurs when all those priorities are considered in a balanced way and the focus of the planning becomes the middle of the model – to develop geographical thinking and not just the delivery of content from the written curriculum. The unit should focus on engaging students in acquiring knowledge that they do not already have, and that they can see as relevant to their lives and engenders a need to know.

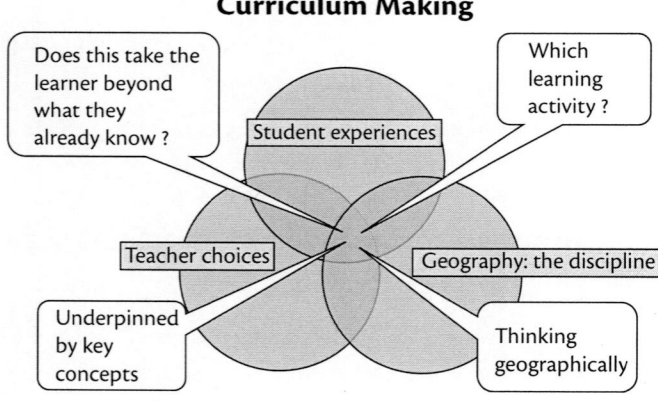

Figure 11.1 Curriculum making model

Source: Lambert & Morgan (2010, p. 50).

Short-answer questions

1. Powerful geographical knowledge does not just include factual content knowledge, but much more. What else does it include?

2. Propose areas of geographical knowledge that you consider powerful knowledge.

Getting started in planning: Identifying the learning requirements

The secondary years (Years 7–10) of the Australian Curriculum: Geography have two units per year level, containing content descriptions for the Knowledge and Understanding and Inquiry and Skills strands. The curriculum also has seven concepts identified to guide conceptual thinking and includes achievement standards to inform assessment.

Figure 11.2 provides a useful visualisation of the deconstruction process employed at this initial stage of the curriculum-making process.

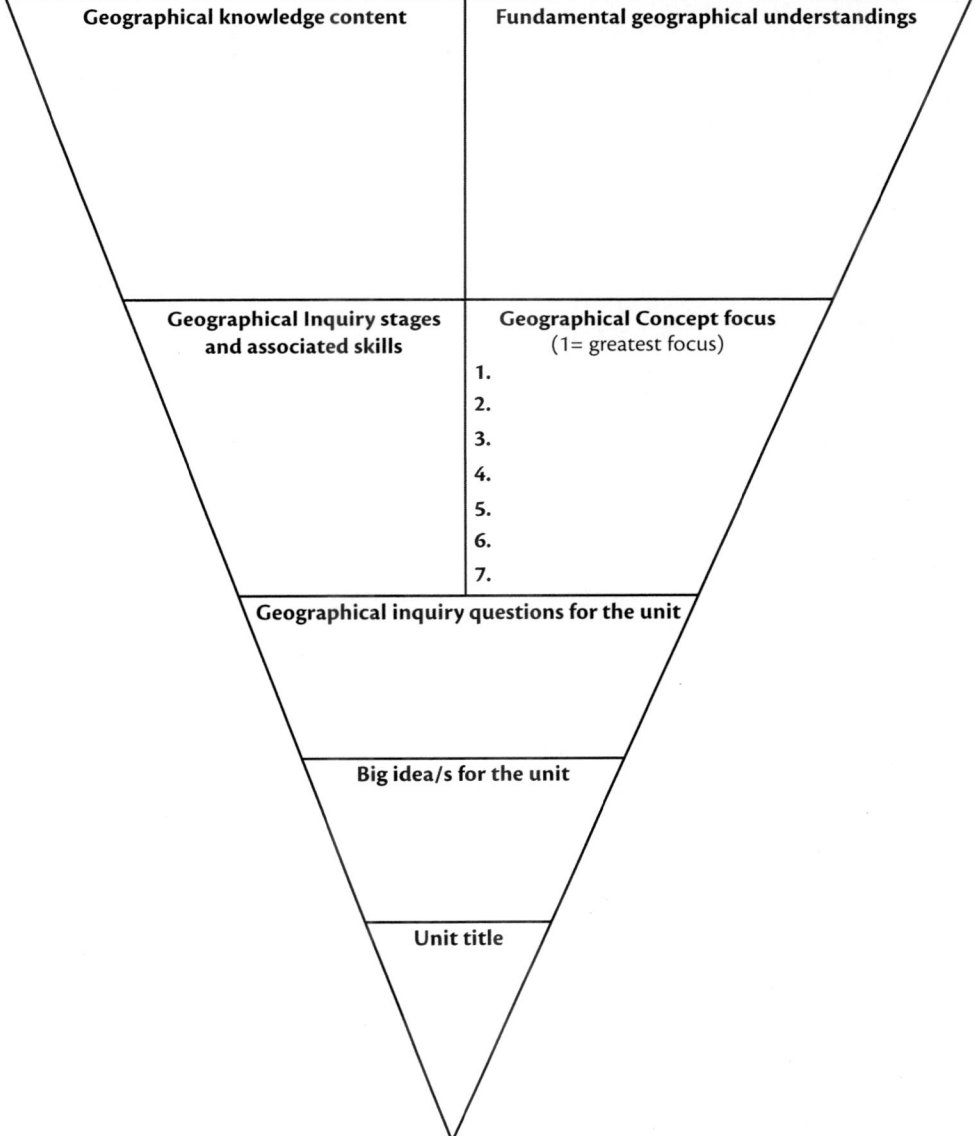

Figure 11.2 Geographical learning requirements curriculum deconstruction triangle

Pause and think

How does the unit deconstruction triangle help in the planning of a unit? Why is it such an important initial curriculum-making activity for the teacher to be guided by an understanding by design approach?

Demonstrating the process of initial planning: Curriculum deconstruction

To demonstrate the deconstruction of a unit, we will focus on the Landforms and Landscapes unit in Year 8 of the Australian Curriculum: Geography.

The Knowledge and Understanding strand identifies the content of the unit as:

- different types of landscapes and their distinctive landform features

- spiritual, aesthetic and cultural value of landforms and landscapes for people, including Aboriginal and Torres Strait Islander peoples

- geomorphic processes that produce landforms, including a case study of at least one landform

- human causes and effects of landscape degradation

- ways of protecting significant landscapes

- causes, impacts and responses to a geomorphological hazard (ACARA 2020).

These content descriptions may be seen as an itemised list, but the initial task of the curriculum-maker is not just to start teaching the content as a checklist, but rather to deconstruct the content descriptions to determine the learning requirements in terms of knowledge, fundamental understandings, skills, conceptual focus and overall purpose, and associated questions.

There are several points to be made before embarking on populating the deconstruction triangle. As with all curriculum planning, the most effective way to populate the template is through discussion with colleagues to clarify the content of the unit, share ideas and provide inspiration for creativity with a big idea and title for the unit of work.

As teachers populate the triangle, they need to be cognisant that the skills section is difficult to complete because the identification of the skills may not be possible until the pedagogy to be employed has been decided during the planning stage of the process – for example, if it is decided that fieldwork will be conducted for the unit, then the field skills will be identified. It is a useful part of the process to be concise when identifying the essential learning requirements of the unit. Being economical with words results in a targeted and precise planning document containing the essentials for planning the unit.

Deconstructing the written curriculum is a creative process of curriculum-making for the teachers with their student cohort in mind. There is no one perfect and correct deconstruction but rather many, depending on the situation of the learning environment it is intended for. The deconstructive interpretation of the written curriculum and, in turn, the details inserted into the template will depend on the geographic skills and knowledge of the teacher/s, the context of the school and community and the experiences and knowledge of the students. As a result, every deconstruction will be unique and fit for purpose as deemed by the teaching professional developing the learning requirements from the written curriculum – the teacher's professional judgement guided by the curriculum.

Figure 11.3 is an example of a group of teachers using the deconstruction triangle to identify the key learning requirements of the Landforms and Landscapes unit of the Australian Curriculum: Geography.

Pause and think

How would you complete the deconstruction triangle for another unit of the Years 7–10 geography curriculum? Where would you start? Have a go at populating the deconstruction triangle for your chosen unit.

Having identified the learning requirements for the unit, the next stage of backward designing when planning a unit is to consider the assessment requirements. This can be done through the Australian Curriculum achievement standards for Year 8 geography or similar assessment outcomes/standards developed by different states for assessment purposes. Before getting down to planning the pedagogy, activities and processes for the unit plan, educators need to explore the nature of quality assessment in geography and ideas about the collection of student evidence of geographic learning.

Assessment

Why and how to assess in geography

Assessment is integral to the planning of teaching and learning, and is a professional responsibility of all teachers. Masters (2014, pp. 1–2) notes that '[t]he fundamental purpose of assessment in education is to establish and understand where students are in an aspect of their learning at the time of the assessment'. When assessing in geography, teachers are required to establish where students are in their learning before, during and after teaching, and to know whether students have learnt what has been taught.

The written curriculum provides standards to support teachers in determining where students are in their learning. In the case of the Australian Curriculum, there are achievement standards to guide assessment. The Australian Curriculum does not provide specified assessment tasks; however, examples of assessment tasks can be accessed in the Work Sample Portfolios on the Australian Curriculum website. The purpose of the Work Samples is to support the achievement standards rather than necessarily providing quality tasks.

Geographical knowledge content

- What are the different types of landscapes?
- What landforms are associated with different landscapes?
- What is an example of a landscape or landform that has a spiritual, aesthetic and cultural value?
- What is geomorphic processes?
- What is an example of a landform created by specific geomorphic processes?
- How do humans negatively impact on landscapes?
- How do landscapes be protected?
- What are geomorphic hazards?
- What are the causes and impacts of geomorphic hazards?
- How can humans mitigate the impact of geomorphic hazards?

Fundamental geographical understandings

- There are different landscapes on Earth, all having identifiable landform features.
- Landscape and landforms have spiritual, aesthetic and cultural valus.
- Landform are the result of a range of geomorphic processes.
- Humans can have a negative impact on landscapes and landforms.
- Humans can intervene to protect landscapes.
- Hazards exist as a result of geomorphic processes.
- Geomorphic hazards have various causes and impacts.
- Humans have a range of ways to respect to and mitigate geomorphic hazards.

Geographical Inquiry stages and associated skills

Questioning
- Develop geographically significant questions.

Representing
- Represent the spatial distribution of a geographical phenomena.

Interpretation
- Interpret geographical data and information
- Apply geographical concepts

Reflecting and responding
- Propose individual and collective responses to a geographical challenge.

Geographical Concept focus
(1 = greatest focus)

1. Environment
2. Place
3. Interconnection
4. Sustainability
5. Change
6. Space
7. Seals

Geographical Inquiry questions for the unit

- How do areas of the Earth differ in geomorphological nature and their suitability for human occupation?
- How do humans perceive and connect with the physical form of the Earth?
- Can humans influence geomorphological processes?
- Are there ways for humans to minimise the impact of geomorphological hazards?

Big idea/s for the unit

Landscape and landforms provide a point of connection between humans and the environment; however, the existence of landscapes and landforms are also threatened or changes by human activity

Unit title

Connection to and perception of the land

Figure 11.3 Draft deconstruction triangle of the content descriptions for the Year 8 Landforms and Landscapes unit of the Australian Curriculum: Geography

Each state is responsible for assessment of the Australian Curriculum and each state has taken the achievement standards and interpreted them differently. Although this chapter references the Australian Curriculum achievement standards, the purpose of the chapter is to stimulate a discussion of assessment rather than to specifically address the achievement standards.

CONNECTION

Key points for teachers

Below are some key points to support teachers to ensure that students have learnt the geography taught:

- Students are able to develop ideas about the geographical changes they know about or experience.

- Teachers develop good questioning skills to assist students to investigate the geographical world and understand the place of close questions and facts in the learning process.

- Students are able to respond to ideas and challenges using precise geographical language in classroom dialogue prior to written responses.

- Students know and understand at the beginning of the unit what the learning intentions are, and therefore what will be assessed.

- Teachers model quality responses in tasks.

- Students are able to focus on the interaction between the physical and the human world – otherwise it is science.

CONNECTION

Guiding points for teachers

Below are some guiding points for teachers associated with quality assessment:

- Engage students in rich conversations after assessing to support them with knowing what they are good at and what they need to get better at.

- Ensure learning is non-competitive between students – the focus should be on progress and improvement.

- Marks and grades can be demoralising for low achievers and not motivational for high achievers. They are generally only useful for 'middle' students.

- Show students how to peer-assess and self-assess as part of the assessment process. Students can help each other with strategies and ideas as well as knowledge.

- Involve students in their learning through feedback. This helps students to have a good knowledge of themselves as learners and their strengths and areas for improvement, and to understand what is expected of them.

- Spending valuable time designing rubrics for 'D' or 'E' grades is not useful. Focus on what success looks like including providing examples and models of above satisfactory responses to the task.

- Consider not providing grades initially when returning marked work to students. Students will be more likely to read written comments if there is no mark or grade on their work. Allow time for students to have a rich conversation with their peers about the comments on their work. Suggest to students that they may like to know their mark or grade at another time.

Pause and think

Describe how you will go about teaching your students the skills they will need to self-assess and peer-assess confidently and competently.

Feedback and observation

While reinforcing the importance of feedback for student learning, various strategies exist for making comments more effective for learning improvement:

- Provide time to re-do and improve after feedback.

- Engage students in conversation rather than just providing a brief written comment.

- Ensure improvement is acknowledged.

- Provide in-class time and supervise rewriting. This is not the place for homework.

At this stage, it is worth considering using marks or grades versus oral or written feedback. Whether we like it or not, assessment can impact students' self-esteem and wellbeing. Assessment should build positive attitudes and self-confidence in young people by assisting them to see what they have achieved and the progress they are making. Engaging students in rich dialogue about their learning leads to improved relationships between the teacher and students, and between students.

Observing students is important and needs to be more than simply watching. Notice what they can do, what their strengths are, how they are developing as learners and how they learn best. To get a clear insight into a student's improvement and progress, the teacher's observation must be non-biased, context free, non-judgemental and factual.

In the documentation, avoid emotive language and drawing conclusions. Instead, focus on what can be noticed – the facts. For example, saying 'needs to try harder and has wasted class time' is not helpful. Instead, a teacher might say 'needs initial support with understanding the task, breaking the task down into manageable'. Also consider pairing this student with another student to help them stay on task.

Follow this process with a reflection on the observations. What can the student do? What else does the student *want* to know? What does the teacher now know about this student? How else might the teacher get more, reliable information about this student?

Designing effective assessment

Teachers ensure effective teaching and monitoring of student learning by being organised and structured in designing assessment. It is important to streamline assessment and to not over-assess. During the design process, teachers will identify the learning intentions, assessment types, strategies and activities, and resources. All aspects of the curriculum are considered, including knowledge content, understandings, skills and conceptual thinking. To do this well, teachers need to be experts in Australian Curriculum: Geography knowledge, concepts, inquiry processes, skills and achievement standards, and to understand the progression in the achievement standard.

Geography provides many opportunities for students to show their learning in authentic tasks, which ask students to relate to real-world issues and solutions. Providing samples of annotated work to demonstrate expectations supports student understanding of expectations and hence success in demonstrating learning.

When designing the assessment, teachers need to think through what is going to be the likely 'best response' from any student and also what will be understood as a satisfactory response. This will ensure that all students can engage with the task and have a chance to be successful.

Working with other teachers when designing tasks assists with ensuring the assessment is valid in terms of content, literacy skills, cultural appropriateness and ability of students. This may be within the school if there is a learning area team, or it may be through a geography teachers' association or online network.

Ensuring quality judgements

Teachers need to be certain that student learning is at the expected standard. Jones and Dineen (2019, p. 275) note that, 'Collaborative moderation is one of the most effective methods of assuring quality judgements, as it strengthens the consistency of educators' judgements.' Some of the best professional learning that teachers can do is with their colleagues discussing student evidence of learning. When students have completed assessment tasks, share this student work with colleagues and discuss whether it shows evidence of aspects of the standard and to what degree it meets the standard. This supports all teachers in reflecting on their practice, sharing ideas, pedagogy and strategies for teaching and learning, developing deeper understanding of the achievement standards and content descriptors as well as confirming their assessment judgements.

Differentiation

Teachers need to ensure assessment activities are accessible to all students so they can demonstrate their proficiency. This requires the assessor to respect student differences in culture or ability when designing assessment activities for students at risk of not achieving or progressing by:

- writing support, including explicit teaching of appropriate language conventions and authorial choices

- ensuring appropriate text complexity when selecting resources and teaching the genre – for example, persuasive writing

- knowing and understanding the assessment standards above and below the year level of the cohort

- selecting interventions and strategies based on ongoing formal and informal assessments of students at risk of not achieving or progressing.

Pause and think

How will you ensure that you are differentiating assessment so all students can access the assessment tasks while still giving all students the opportunity to meet the relevant achievement standard?

The place of collaborating

When providing opportunities for students to work together, share the resources between the students rather than dividing the tasks among the students. This sets up a group dependency model among the students for solving an issue or following an inquiry. When assessing collaborative student work, focus on the thinking they have engaged in, not just the product they have produced. Ask questions about their individual participation to gather further assessment data on cooperation and collaborative problem-solving as described in the personal and social capability evident in the Australian Curriculum.

After reviewing the learning requirements of the Landscape and Landforms unit, Table 11.1 identifies the aspects of the Achievement Standards that can be addressed in the unit plan.

Table 11.1 Identified assessment aspects for the Landforms and Landscapes unit from the Year 8 achievement standard

Geographical Knowledge and Understanding	Geographical Inquiry and Skills	
	Interpret and analyse	**Communicate**
Explain geomorphic processes that produce landforms		**Use** geographical terminology
Explain interconnections of human causes and effects of landscape degradation – floods	**Propose** action on ways of protecting significant landscapes	**Present** findings, arguments and ideas using labelled sketches, diagrams, photographs, maps
	Explain spatial distributions and patterns	**Use** digital or non-digital forms of communication to **propose** explanations and **predict** outcomes
	Analyse maps, data **Evaluate** sources	

Pause and think

How would you go about completing a similar table for another unit of the Years 7–10 geography curriculum? Where would you start? Have a go at populating the table for your chosen unit.

Having determined the aspects of the achievement standard to be assessed for the unit, tasks that will provide evidence of learning can be planned. Avoid giving students a long-term, complex task at the outset. Scaffold the assessment with the development of tasks that will lead to the final presentation, which is a proposed action and outcome for a geographical challenge.

Table 11.2 is a suggested plan for the assessment tasks for the Landforms and Landscapes unit. These tasks are all closely linked with the geographical inquiry

Table 11.2 A suggested plan for the assessment tasks for the Landforms and Landscapes unit

Assessment task	Evidence
1. **Annotated visuals** (formative, Weeks 1–2)	Source and annotate photographs or annotate sketches to identify the difference between landscapes and landforms overall and those of the Country where the school is located. Provide infographic or labelled drawing of a geomorphic process. Source and annotate photographs OR produce an annotated model to demonstrate an explanation of a geomorphic process.
2. **Map reading and production** (formative, Weeks 1–2)	Represent spatial distribution of the landscapes and landforms of Country using large- and small-scale maps. Use the map to explain and locate Country and the spatial patterns of landscapes and landforms across the Country where the school is located.
3. **Artwork, song and story analysis** (formative, Weeks 3–4)	Select and evaluate sources that link Aboriginal and Torres Strait Islander artwork, song, story and culture to significant landscapes and landforms in Australia and on the Country where the school is located.
4. **Quiz** (formative, Weeks 1–8)	Conduct a weekly kahoot-style quiz focused on key aspects of learning as appropriate to the week, such as terminology and underpinning concepts.
5. **Development of a geographical question** to frame a research task, and conduct the research (formative, Weeks 3–4 and Weeks 5–6)	Write a one-page response and/or develop an infographic to explain the possibilities for protecting and managing the landscapes and landforms of the Country where the school is located in response to processes of change.

Table 11.2 (*cont.*)

Assessment task	Evidence
6. **Fieldwork** (teacher discretion – can be part of a cultural immersion day on Country)	Produce a field report if a field trip is possible.
7. **Case study** (summative, for presentation in Weeks 7–8)	Produce a three-minute oral presentation that synthesises learning about *Connection to Country and perception of the land: Landscapes and landforms* and proposes possible responses to sustainably protect and manage the landscapes and landforms of Country in response to the challenges of change and value placed upon the environment.

questions and the big ideas in the deconstruction triangle described earlier in this chapter (see Figure 11.3 above). Early in the unit, students will be asked to select a location in Australia that will be the focus of their study. If possible, include a field trip to a location of interest. Provide opportunities for students to negotiate both the place and the way they present their findings.

This part of the planning process has highlighted that the effective use of assessment by educators and learners is an integral part of the planning of learning and teaching, to ensure students maximise their successes and achievements. Assessment approaches that 'allow learners to demonstrate their knowledge and understanding, skills, attributes and capabilities in different contexts, are key to our understanding of the progress learners are making' (Education Scotland 2020, p. 4).

Short-answer questions

1. What do I want students to be able to know, do and understand by the end of the unit of work that is distinctly geographical?

2. How will I know when students have achieved this or are working towards its achievement?

3. How will the students know when they have achieved this or are working towards its achievement?

4. How can you ensure consistency and confidence in your own assessment judgements?

Unit planning

The identification of the learning requirements and relevant assessment standards at the beginning of planning is critical for the planning process. Understanding by design refers to this approach as having the 'end in mind' as a unit of work is planned.

Most importantly, as the unit plan is developed the teacher designing the program needs to ask the following questions:

- Does the unit reflect the essence of the intended learning embodied in the curriculum?

- What powerful geographical knowledge does the unit contain? Why should it be taught? Is the learning achievable for all students?

- Does the assessment reflect the standards outlined in the curriculum?

- Are the activities and tasks age-appropriate?

- Are the activities and processes of the plan creative to engage the students?

- Has a variety of teaching methods and approaches been used to cater for all learning styles in the group?

- Will the activities and processes be seen by the students as relevant to their world and useful for their future as active and informed citizens?

- Do the assessment tasks provide opportunities for the students to show evidence of their learning in accordance with the identified assessment standards for the unit?

What would having the 'end in mind' look like for a unit plan for Landforms and Landscapes in Year 8? Earlier in the chapter, it was called 'front-ending' the targeted knowledge, understanding, skills and conceptual understanding for the unit. A detailed sample eight-week unit plan for Landforms and Landscapes, as well as a sample weekly plan, can be found on the website for this text.

Conclusion

Quality teaching and learning do not just happen! As with designing a house or event, an important role of a teacher is to design learning. In recent years, the role of teachers as designers has become increasingly evident, as the need to develop effective and relevant teaching programs has increased due to the constantly changing curriculum and community demands for quality education. This chapter has explored a process for planning in geography that involves being clear about what the purpose and aims of the teaching program are from the very beginning.

The planning process involves:

1. developing curriculum-relevant and contextualised learning requirements from the written curriculum

2. identifying the relevant assessment standards for the unit and developing assessment tasks to provide evidence of learning

3. planning the unit in terms of learning tasks, assessment tasks, resources, teaching materials and programming.

Developing a relevant, creative, achievable and engaging unit plan for students is at the core of quality teaching, and it requires time and effort to achieve. Ideally, such processes are most effective, creative and enjoyable when undertaken with colleagues. The best teaching occurs when a teacher, either individually or collaboratively, designs a unit of work and does not just pick up something developed by someone else. In fact, the processes of intellectualisation, creating and personalising of the curriculum by teachers as curriculum-makers is fundamental to quality education. Units of work, contextualised for students and rewarding to teach, will not just happen but rather need to be a conscious and concerted process by teachers. Hopefully, the planning ideas in this chapter will provide teachers with the guidance, processes and inspiration to be curriculum-makers and not just curriculum receivers and deliverers.

Bringing it together

1. Understanding by design (UBD) is an approach to unit planning that 'front-ends' the desired outcomes of the learning. Why, with such an approach, is it important to identify the learning requirements of the unit as the first stage of the process?

2. Briefly describe the three UBD stages.

3. What makes knowledge powerful in geography?

4. Curriculum-making is a way of looking at the work of the teacher when they are developing their courses and approaches. Describe the model and provide a considered view on its relevance and usefulness to planning.

5. What do you understand as curriculum deconstruction and how would you go about populating the triangle template? Where would you start – with the big idea or the breakdown of the content descriptions?

6. What key ways will you use to support your students to have a key understanding of their strengths and areas for improvement?

7. How will you use all the assessment data you collect to identify individual students' needs and plan learning?

References

Australian Curriculum, Assessment and Reporting Authority [ACARA] (2020) *The Australian Curriculum Version 8.4,* Canberra: ACARA, retrieved from www.australiancurriculum.edu.au/f-10-curriculum/humanities-and-social-sciences/geography

Education Scotland (2020) *Assessment within the broad general education: A thematic inspection*, Edinburgh: Education Scotland, retrieved from https://education.gov.scot/media/plwh4m3d/assessment-within-bge.pdf

Jones, S. & Dineen, C. (2019) Effective assessment practices, in D. Green & D. Price, eds, *Making humanities and social sciences come alive*, Melbourne: Cambridge University Press, pp. 263–78.

Lambert, D. (2015) National Curriculum: a possibilist interpretation, retrieved from https://slideplayer.com/slide/8717665

—— (2016a) Explaining curriculum making, retrieved from www.youtube.com/watch?v=q-VtcTa1Ypc

—— (2016b) Curriculum making in geography education, retrieved from www.ucl.ac.uk/ioe/case-studies/2016/jan/curriculum-making-geography-education

Lambert, D. & Morgan, J. (2010) *Teaching geography 11–18: A conceptual approach*, Maidenhead: Open University Press.

Masters, G. (2014) Assessment: Getting to the essence, *Designing the Future*, 1, retrieved from www.acer.org/files/uploads/Assessment_Getting_to_the_essence.pdf

Wiggins, G. & McTighe, J. (2005) *Understanding by design*, Alexandria, VA: ASDC Books.

Young, M. (2013) On the powers of powerful knowledge, *Knowledge and the Future of the Curriculum*, 1(3), 229–50.

The professionally engaged geography teacher

Susan Caldis and Alaric Maude

Learning objectives

By the end of this chapter, you should be able to:

- communicate a broad understanding of the nature and importance of professional engagement

- develop an understanding of how professional engagement can occur in geography

- identify achievable strategies for professional engagement to build and maintain professional capacity as a teacher of geography

Introduction

Professional engagement occurs when teachers actively participate in a range of professional learning opportunities. The chosen professional learning opportunities should respond to self-identified learning goals and desired areas of growth. Professional engagement is characterised by the sustained and proactive demonstration of respectful, professional interactions with students, colleagues, parents/carers and members of the community. For teachers of geography, professional standards are identified in the Professional Standards for the Accomplished Teaching of School Geography (Hutchinson & Kriewaldt 2010; Kriewaldt & Mulcahy 2010). More broadly, professional engagement is identified as one of the three domains of teaching, and is evident in the Australian Professional Standards for Teachers (AITSL 2017). It is also necessary for teachers to demonstrate and collect evidence about professional engagement for accreditation purposes.

This chapter explores what professional engagement as a geography teacher looks like by identifying several geography-specific communities of practice and suggesting a range of resources suitable for teacher professional learning (TPL) in geography. Attention is then focused on the nature of professional engagement overall.

Throughout the chapter, the reader is challenged to reflect on and consider how they can continue to maintain and build their capacity as a graduate and proficient early career teacher of geography.

What is professional engagement?

Pause and think

1. In 50 words or less, outline your current understanding of the nature and importance of professional engagement overall.

2. Identify up to three areas of teaching practice that you would like to continue to learn about as a graduate or proficient early-career teacher of geography.

3. Outline up to three ways in which you anticipate being able to professionally engage with the geography teaching community.

Professional engagement occurs when teachers can identify their own learning needs and know how to access professional learning opportunities to meet them. Valuing 'opportunities to engage with communities within and beyond the classroom' in order to enrich the learning of both students and oneself is another important part of professional engagement (AITSL 2017). The enactment of professional engagement therefore occurs when the practitioner can integrate and apply learning from the domains of professional knowledge and professional practice to create a teaching environment where learning is valued, and where the teacher takes an active role in their own learning (AITSL 2017).

Professional engagement is one of the three domains of the Australian Professional Standards for Teachers; the other two domains are professional knowledge and professional practice. There are two Standards in the professional engagement domain:

- Standard 6: Engage in professional learning

- Standard 7: Engage professionally with colleagues, parents/carers and the community.

Each Standard has four sub-sections, as shown in Table 12.1.

For the remainder of the chapter, the focus is on developing capacity for professional engagement in the context of teaching geography.

Professional engagement for teachers of geography

The main purpose of professional engagement in geography is to strengthen and extend teacher understanding about evidence-informed developments in research and content,

Table 12.1 Standards and their sub-sections related to the professional engagement domain

Standard 6 Engage in professional learning	Standard 7 Engage professionally with colleagues, parents/carers and the community
6.1 Identify and plan professional learning needs	7.1 Meet professional ethics and responsibilities
6.2 Engage in professional learning	7.2 Comply with legislative, administrative and organisational requirements
6.3 Engage with colleagues to improve practice	7.3 Engage with the parents/carers
6.4 Apply professional learning to improve student learning	7.4 Engage with professional teaching networks and broader community

Source: AITSL (2017).

pedagogy, curriculum and assessment. Another purpose is to be able to effectively demonstrate how the knowledge, skills and capabilities of geography apply to everyday life. In so doing, a geography teacher is able to communicate the distinctiveness and rigour of the subject in a school setting to students, colleagues and parents/carers. By engaging with the distinctiveness and application of geography as a subject, teachers are also able to develop their understanding of the profession of geography – what the subject looks like outside school. They can introduce students to 'professional geographers', such as urban planners. Such introductions, in real life, in text or by virtual means, will enable students to engage in discussion with their teachers about the relevance of geography. Students will also be able to develop their own ideas about further study and career pathways associated with the study of geography. It is important for both teachers and students of geography to have a clear understanding of the relevance of geography and its representation in higher education, industry and the community.

Professional engagement in geography can occur through activities such as:

- joining the state- or territory-based professional association where you live and/or teach – for example, the Geography Teachers' Association of South Australia (GTASA)
- joining a public geography society, such as the Royal Geographical Society of Queensland (RGSQ)
- attending and presenting at national conferences (e.g. the Australian Geography Teachers Association biennial conference) or state conferences (e.g. the annual conference of the state- or territory-based association for geography teachers), webinars or other professional learning- and networking-focused events such as academic reading groups

- embarking upon geography-focused participatory action research projects with an academic partner
- actively contributing to a distinct community group (online or face to face) connected to the subject, which is not necessarily restricted to teachers (e.g. bush regeneration).

For guidance around professional engagement in a geography teaching context, practitioners are encouraged to familiarise themselves with the Professional Standards for the Accomplished Teaching of School Geography, otherwise known as the GEOGstandards. The GEOGstandards are the outcome of an empirical research project focused on the teaching practice of specialist, experienced geography teachers from schools across New South Wales, South Australia and Victoria (Hutchinson & Kriewaldt 2010; Kriewaldt & Mulcahy 2010). The GEOGstandards demonstrate the common and distinctive elements of exemplary pedagogical and professional practice in the teaching of geography, and are intended for use as a tool of self-reflection and self-assessment about teaching practice in Geography. Table 12.2 sets out the nine Standards (revisit Chapter 8 for further discussion).

Table 12.2 The Professional Standards for the Accomplished Teaching of School Geography

GEOGstandard
1. Knowing geography and the geography curriculum
2. Fostering geographical inquiry and fieldwork
3. Developing geographical thinking and communication
4. Understanding students and their communities
5. Establishing a safe, supportive and intellectually challenging learning environment
6. Understanding geography teaching – pedagogical practices
7. Planning assessing and reporting
8. Progressing professional growth and development
9. Learning and working collegially

Source: Hutchinson & Kriewaldt (2010); Kriewaldt & Mulcahy (2010); GEOGstandards (2010).

Pause and think

1. Retrieve the 50 words or less statement from question 1 in the previous 'Pause and think'. Highlight something that was confirmed and/or challenged after reading through sections 1 and 2. Now add up to 50 words to the statement to reflect the

development in your understanding about the nature and importance of professional engagement in geography.

2. With reference to your responses for questions 2 and 3 in the previous 'Pause and think', highlight the area(s) that remain something on which you wish to focus for teaching practice and potential engagement with the geography teaching community. Add another area or areas for teaching practice and professional engagement if necessary. Identify the relevant GEOGstandard(s) and sub-section(s) for each one.

Using the GEOGstandards to enact professional engagement in the teaching of geography

The first, eighth and ninth GEOGstandards relate most directly to professional engagement in a geography teaching context.

GEOGstandard 1: 'Knowing geography and the geography curriculum' focuses on developing a deep and applied understanding of geography as a school subject and as a discipline. Within this standard, teachers are expected to work towards and demonstrate:

- knowledge about the 'breadth and depth of the academic discipline, including its concepts, skills, values and understanding'

- the ability to assist students in developing their understanding that geography is a distinctive subject that links the 'physical sciences, the social sciences and the humanities'

- understanding about 'current curriculum documents and reasons for curriculum change'

- understanding about the location of geography in a 'wider educational context and make connections with other curricular and co-curricular areas' (GEOGstandards 2010, p. 3).

As a discipline, geography spans the sciences, social sciences and humanities, but as a school subject it is typically identified in the key learning area of the humanities and social sciences (HASS) or studies of society and environment. Such identification can diminish understanding about the interdisciplinary nature of the subject and can be a barrier to recognition of geography as a contributor to the field of science, technology, engineering and mathematics (STEM) (Caldis 2019; Caldis & Kleeman 2019a; NCGS 2018). Geography as a subject in schools also has one of the highest proportions of out-of-field teachers involved in teaching the subject. The implications of such a situation include concerns about the quality of instruction compared with geography classes being taught by a specialist teacher, because subject content knowledge is a vital component for the effective teaching of geography (Caldis 2017; Caldis & Kleeman 2019b; Lane 2009; NCGS 2018). To have a strong command of geographical content knowledge, together with a clear understanding of the discipline-specific pedagogies, enables specialist geography teachers to provide a thorough, connected learning experience where theoretical processes are applied to real-life examples and lived experience (Caldis & Kleeman, 2019b; Shreeve 2018). Furthermore, geography teachers who know their content and understand the curriculum are better able to plan for and recognise a progression in geographical learning; they are also more able to recognise when students demonstrate misconceptions about geographical processes (Lane 2015).

To professionally engage with GEOGstandard 1, geography teachers are encouraged to continue developing their content knowledge from academic research occurring across the discipline. For example, joining the Institute of Australian Geographers (IAG) provides access to their peer-reviewed journal, *Geographical Research*, a reduced fee for the annual conference and an opportunity to attend member-only events such as webinars where researchers share key findings from their most recent work. Another strategy is to seek the development of partnerships with universities to collaboratively design and implement participatory action research projects focused on an area of interest, such as discipline-specific pedagogical practice (Lane & Caldis 2018) To join an international professional association for geography teachers, such as the Geographical Association (www.geography.org.uk) in the United Kingdom could also be well worthwhile. There are many synergies between the Australian Curriculum: Geography and the Geography curriculum in the United Kingdom. The Geographical Association produces a wide range of resources and also publishes three journals: *Primary Geography* (for early years and primary teachers); *Teaching Geography* (for secondary school teachers); and *Geography,* an international journal for lecturers, teachers and senior secondary students. Another way to professionally engage with content knowledge from the field and connect it with curriculum-related content for the Geography classroom is the City Road podcast (see Connection box).

CONNECTION

Using the City Road podcast to develop professional engagement with geography in higher education and explore Place and Liveability

The City Road podcast (www.cityroadpod.org) is a portal of evidence-informed podcasts about cities, urban life and urban-planning-related issues. In March 2020, the founder of City Road podcast, Dr Rogers, reached out via social media seeking contributors to what would become a rapidly mobilised international podcast project of collective biographies and digital storytelling. The project is called 'Listening to the city in a global pandemic', and it aims to provide an open-access platform of resources to help people better understand life in the COVID-19 city.

Four themes are used to interpret the observations of place and time in a suburb, city or village:

- **Visibility and invisibility:** more aware of the disease as being invisible yet manifested through its visible traits such as masks

- **Privilege and privation:** those who can self-isolate in private space compared to those who are marginalised, homeless or living in communal accommodation

- **Selfishness and solidarity:** social and personal behaviours such as hoarding vs sharing, caring messages for community from community

- **Absence and presence:** emptiness, who and what is here or not here? 'less buzz' metaphorically and literally in a city.

The Australian Curriculum Year 7 unit, Place and Liveability, has two content descriptions where several podcasts from 'Listening to the city in a global pandemic' can be used to prompt thinking about place, liveability, personal experience and a future-focused approach to managing or adapting to life in a city through the four themes:

- the influence of accessibility to services and facilities on the liveability of places

- the influence of environmental quality on the liveability of places.

The podcasts from the City Road podcast provide an opportunity to consider new ways of covering Place and Liveability to reflect the current time; to do so would also encourage students to apply illustrative living and dynamic examples from in the field and in the moment to fairly static, well-known and rehearsed case studies. It is also worth emphasising the power of observation and audio-recording as a fieldwork tool. It may be the case that audio-recordings of observations in place about seeing, hearing, smelling and feeling, interpreted through such themes of absence and presence, may start to form an integral part of future fieldwork for Place and Liveability.

'Listening to the city in a global pandemic' provides a unique and accessible pool of resources suitable for use in the school-based geography classroom. As geography educators, we all have a role to play in emphasising the distinctiveness, relevance and visibility of our subject. One way we can do this is to make a concerted effort to use and connect current research happening in the field to geographical learning from the syllabus (Caldis 2018, 2020).

By reflecting on and demonstrating capacity in GEOGstandard 1, geography teachers are also able to collect evidence for accreditation against Standard 2: 'Know the content and how to teach it' from the Professional Knowledge domain of the Australian Professional Standards for Teachers (AITSL 2017).

Another important aspect of professional engagement is to participate in and contribute to teacher professional learning about geography teaching, as described in GEOGstandard 8 'Progressing professional growth and development' and GEOGstandard 9 'Learning and working collegially' (see Table 12.3).

Table 12.3 GEOGstandards and their sub-sections related to professional engagement and professional learning in a geography teaching context

GEOGstandard 8 Progressing professional growth and development	GEOGstandard 9 Learning and working collegially
8.1 Continue to learn and develop as teachers, acknowledging that the greater the teacher learning the more students learn as well	9.1 Build a culture of professional improvement by learning from and with fellow teachers as well as learning from research, supported by the professional and school community; create conditions for teachers to teach each other, support their peers and deepen their knowledge about students and subject
8.2 Recognise that the subject of geography is dynamic and evolving, therefore seek opportunities to further develop a disciplinary knowledge base	9.2 Play an active role in professional associations, promote professional learning and talk publicly about practice
8.3 Commit to formal learning and critical reflection on experiences both within the classroom and more widely through travel literature, the arts and through engaging with professional learning communities	9.3 Distribute and share teaching expertise towards the continuing construction of a professional knowledge base for school geography
	9.4 Communicate educational ideas and promote geographical education towards contributing to the resilience and renewal of their professional field

Sources: Hutchinson & Kriewaldt (2010); Kriewaldt & Mulcahy (2010); GEOGstandards (2010)

The Australian Geography Teachers Association (AGTA) is the peak body for geography teachers, representing members of the affiliate associations across the country.

- Geography and History Teachers' Association of the Northern Territory
- Geography Teachers Association of NSW & ACT Inc.
- Geography Teachers' Association of Queensland

- Geography Teachers' Association of South Australia

- Geography Teachers' Association of Victoria

- Geographical Association of Western Australia

- Tasmanian Geography Teachers' Association.

In addition to joining the teacher-focused geography associations, it is also possible to join the academic or public geography society, to learn more about geography in higher education and in the community generally. These societies are:

- Geographical Society of NSW

- Institute of Australian Geographers

- Royal Geographical Society of Queensland

- Royal Geographical Society of South Australia.

Participating in and contributing to online and in-person professional networks for the relevant state or territory where one resides are essential to professional growth. Online professional network contributions will typically occur through engaging with social media (international and Australian focus) – for example:

- Twitter: #geographyteacher; @GAWA; @GTAV; @GTANSWACT; @GeographicalResearch; geography departments of universities (e.g. @ACCESS_GEOG @geoplanMQ); Geography associations from overseas (e.g. @The_GA, @RGS_IBG, @theAAG, @CanGeoEdu @CanGeographers @RCGS_SGRC @SDGoals @TheGlobal Goals @AsiaEducation)

- Facebook: Community of Geography Teachers Australia; Facebook pages from the Geography Teachers' Associations around Australia and internationally; National Council for Geography Education.

Becoming a member of an association or society enables active participation in a geographical community. Membership provides access to resources and publications. It also provides opportunities to network and to attend events such as annual conferences, lectures, workshops and tours. By joining an association or society and attending and contributing to organised events, geography teachers work towards achieving GEOGstandards 8 and 9 in the context of the Australian Professional Standards for Teachers (AITSL 2017): Standard 6: 'Engage in professional learning' and Standard 7: 'Engage professionally with colleagues, parents/carers and the community'.

Professional engagement is part of a professional learning process. Professional learning is about developing and changing practice in response to teaching context, personal experience, beliefs and values, and as part of a wider interconnected system (Timperley et al. 2020). Once identification of professional learning needs occurs, the teacher acts to find suitable professional learning opportunities. Professional learning activities can occur formally and informally, however, they should be aligned with school improvement priorities and goals, strategies for improving student learning outcomes and ways of teaching specific curriculum content (Edge, Reynolds & O'Toole 2015).

Professional learning also needs to be ongoing within considered reflection about how such learning is relevant and enables transformative teaching practice to occur (Hill & Jones 2018). Professional learning initiatives related to developing understanding about geographical content and discipline-specific pedagogical practice will often occur through participation in courses offered by professional associations, or by joining professional learning communities such as a local school or regional network of geography educators and/or through exploring academic partnerships with universities. A professional learning community is a learning-oriented group that focuses on understanding and developing ways to promote the professional growth of teachers and learning capacity of students. Subject knowledge, curriculum and student learning are the focus of attention, and a professional learning community can occur within or outside of a school. Those in the community work collaboratively to share ideas and interrogate teaching practice in alignment with the professional teaching standards (AITSL 2017).

CONNECTION

What it could look like to be a professionally engaged geography teacher

There are many educational and teaching-focused aspects a teacher of geography will need to engage with and fulfil from the Australian Professional Standards for Teachers [APST] (AITSL 2017). However, it is important to remember the importance of having a subject-specific lens to apply to aspects of professional engagement and teacher professional learning and engagement.

For example, a teacher of geography may work towards meeting requirements for APST Standard 6 'Engage in professional learning' (AITSL 2017) by including a subject-specific item in their overall plan for professional learning needs. To gain accreditation for APST Standard 6, a geography teacher would be wise to use GEOGstandards 1, 8 and 9 for guidance about how to engage with the subject and engage with professional learning specific to geography. For example, a geography teacher may wish to upskill in the use and application of geospatial technologies, or to learn more about the use, gathering and interpretation of statistics. Once the professional learning need(s) are identified, part of the plan to meet such needs will be likely relate to APST Standard 6 focus area 6.1 and 6.2 and also APST Standard 7 'Engage professionally with colleague, parents, carers and the community' (AITSL 2017).

To follow through with the example for APST Standard 6, focus area 6.1 and 6.2, the plan might be for the geography teacher to complete an accredited session about the nature and use of geospatial technologies with either the professional association or an equivalent provider. The accredited session could be part of an overall conference program or be a designated course such as a one-day intensive workshop. Such a session could occur online, face to face or in hybrid form. Alternatively, engagement with professional learning can also occur through the identification and completion of peer-reviewed professional readings with the development of annotations to form an evidence base of engagement with professional learning. Such readings may include journal articles or chapters in textbooks. The act of participation in and engagement with such professional learning activities will therefore enable evidence for Standard 6.2 to be generated. Upon completion of professional learning, the act of applying and sharing the learning needs to occur, which leads to the focus areas of 6.3 and 6.4 in APST Standard 6; it may also provide an opportunity to connect with APST Standard 7, focus area 7.4.

As the geography teacher completes an accredited session about geospatial technologies, or completes a series of professional readings about statistics, it is then appropriate for this teacher to share learning with colleagues who are also teaching geography and may require assistance, and therefore improve their practice in such an area (APST Standard 6 focus area 6.3). The sharing could occur during a regular faculty or stage-based staff meeting, or be a separate professional learning session. Part of sharing could include some possible strategies suitable for use in the classroom. The use and application of geospatial technologies and/or statistical analysis can occur across a range of units in the Australian Curriculum: Geography, but may suit a more sustained focus in, for example, Place and Liveability, or the Geographies of Human Wellbeing (ACARA 2015). After engaging with colleagues to share learning with the intent of improving practice, the time comes to apply such learning in the classroom and to monitor student progression in learning about their use and application of geospatial technologies and/or statistics. (To capture evidence about improvement in student learning, please see Chapter 11.)

Pause and think

1. Retrieve the 50 words or less statement from question 1 in the earlier 'Pause and think'. Highlight something that was confirmed and/or challenged after reading through the chapter so far. Now add up to 50 words to the statement to reflect the development in your understanding about the nature and importance of professional engagement in geography.

2. With reference to your responses for questions 2 and 3 in the earlier 'Pause and think', asterisk the highlighted area(s) that remain as something on which you wish to focus for teaching practice and potential engagement with the geography teaching community. Add another area(s) for teaching practice and professional engagement together with the relevant GEOGstandard(s) and sub-section(s) as

appropriate. Annotate the responses as to whether or not there is an extension of understanding about possibilities for engagement or if there is an amendment/ adjustment to be made as a result of reading through the chapter so far.

Using geographical resources to professionally engage with geography in the classroom

It is important for geography teachers to engage with a range of geographical tools (such as statistics) and resources (such as textbooks) to effectively share and communicate geographically distinctive data and information with and to students. This section provides an overview of popular geographical tools that are suitable either for use in the classroom with students or as part of professional growth and development. A list of useful resources can be found on the website for this text.

Geographical tools

Why are statistics important and how can they be accessed and used in geography? A growing number of online sources are available for global and national statistics. Some present data as maps and time series graphs, and some are animated and show change unfolding over time. Some have free software that students can use to make their own visualisations. The following is a shortlist of useful sources:

- **Gapminder:** Gapminder has interactive and animated graphs for a wide range of statistical indicators for countries over many years, as well as videos, teachers' guides, games and presentations.

- **Statsilk:** Statsilk is an Australian company based in Sydney, which produces a range of tools for mapping and graphing world statistics, some of them freely accessible. The two main ones are:

  - StatPlanet World Bank

  - StatTrends.

- **Australian Bureau of Statistics (ABS):** The ABS provides a very wide range of demographic, economic, social, health and environmental statistical information. It is an essential source for data on your suburb, local government area, town, city or state/territory. Many of the statistics can be downloaded as Excel spreadsheets with which students can work, so look at the Downloads tabs. Some of the most useful ABS products are:

  - QuickStats

  - Data by region

- Catalogue Number 3101.0 – Australian Demographic Statistics

- Catalogue Number 3218.0 – Regional Population Growth, Australia

- Catalogue Number 3412.0 – Migration, Australia.

Professionally engaging with geography beyond the classroom

Many students, parents/caregivers, community members and colleagues within the school will ask geography teachers about areas of further study and employment related to the study of geography.

A major in geography leads to a 'content-focused' career specialisation and/or a 'skills-application' focused pathway. The first often leads to employment in 'environmental management, environmental policy, natural hazard management, urban planning, regional development' and tourism, although some may require further study for a professional qualification (e.g. urban planning and teaching) (NCGS 2018, p. 11).

A 'skills and capabilities' focused career pathway includes employment in areas where the ability to comprehend information and synthesise ideas from the natural and human sciences is important, together with knowledge of the world and its diversity of places, environments and peoples (NCGS 2018). Such capabilities are variously described as 'soft, enterprise or transferable skills, and are increasingly sought by employers' (NCGS 2018, p. 11). Those gained from the study of geography include:

- breadth of thinking

- the ability to see interconnections

- teamwork skills

- analytical skills

- digital literacy

- oral, written and visual communication skills

- problem-solving

- problem identification

- interviewing

- strategic thinking

- critical thinking (NCGS 2018, p. 11).

These are also the capabilities needed by young people to successfully navigate their way through a series of careers in different occupations and industries over their working life, which is 'rapidly becoming the career path of the future' (NCGS 2018, p. 11). Both the knowledge and the capabilities gained from a study of geography are valued in a wide variety of careers (NCGS 2018). For example, in the United Kingdom,

> data suggest that many Geography graduates get jobs in banking, finance, marketing and other types of business. In fact, one advantage of a Geography degree is that it does not have a set career path but can lead in many directions. (NCGS, 2018, p. 11)

Conclusion

When geography teachers remain professionally engaged in a subject-specific way, they are able to:

- assist students and colleagues in developing their understanding of geography and its distinctive ways of thinking and finding out

- understand the depth and breadth of the discipline and the role of geography in a broader educational context to make connections with other curriculum and learning areas. In so doing, professionally engaged geography teachers are able to identify the contribution of and opportunity for geographical learning in cross-faculty project-based learning units of work, for example in STEM.

- contribute to contemporary debate and professional learning initiatives about curriculum documents and the reasons for curriculum change

- develop and communicate a clear understanding about the contribution of geography to student learning and about what will be gained by students from their study of geography – in short, you become the subject ambassador because you have found your own 'buy-in' to the subject and can communicate its value in a meaningful way.

Bringing it together

1. Why should teachers become professionally engaged?

2. If you had to choose one type of professional engagement on which to focus for the next year, what would it be and why?

3. What do you think is the value of having professional standards for geography teachers?

4. The chapter has a list of associations and societies that a geography teacher could join. If you had to choose one of these, which one would it be and why?

5. The chapter recommends that teachers engage in participatory action research. What would you choose to research and why?

References

Australian Curriculum Assessment and Reporting Authority [ACARA] (2015) *Foundation to Year 10 Australian Curriculum: Geography*, Canberra: ACARA.

Australian Institute for Teaching and School Leadership [AITSL] (2017) *Australian Professional Standards for Teachers*, Canberra: Education Services Australia.

Caldis, S (2017) Teaching out-of-field: Teachers having to know what they do not know, *Geography Bulletin*, 49(1), 13–17.

——(2018) Using podcasts and journal articles as a tool of professional learning and a tool of instruction in the Stage 6 geography classroom, *Geography Bulletin*, HSC Special Edition, 1, 26–32.

——(2019) STEM or HASS: Where is geography enabled or constrained?, *Professional Educator*, 20(1), 36–40.

——(2020) Teaching in the COVID-19 city; Listening to the COVID-19 city, *Geography Bulletin*, 3, 93–6.

Caldis, S. & Kleeman, G. (2019a) Geography and STEM, *Geographical Education*, 32, 5–10.

——(2019b) Out of field teaching in geography, *Geographical Education*, 32, 11–14.

Edge, K., Reynolds, R. & O'Toole, M. (2015) Contextual complexity: The professional learning experiences of seven classroom teachers when engaged in 'quality teaching', *Cogent Education*, 2, doi:10.1080/2331186X.2015.1120002

GEOGstandards (2010) *Professional Standards for Accomplished Teaching of School Geography*, retrieved from www.agta.asn.au/files/Professional%20Standards/geogstandards.pdf

Hill, J. & Jones, M. (2018) Professional development, in M. Jones, ed., *The handbook of secondary geography*, Sheffield: Geographical Association.

Hutchinson, N. & Kriewaldt, J. (2010) Developing geography standards: Articulating the complexity of accomplished geography teaching, *Geographical Education*, 23, 32–40.

Kriewaldt, J. & Mulcahy, D. (2010) Professional standards for teaching school geography, *Curriculum & Leadership Journal*, 8(20), retrieved from http://cmslive.curriculum.edu.au/leader/professional_standards_for_teaching_school_geograp,31904.html?issueID=12165

Lane, R. (2009) Articulating the pedagogical content knowledge of accomplished geography teachers, *Geographical Education*, 22, 40–50.

——(2015) Primary geography in Australia: Pre-service geography teachers' understandings of weather and climate, *Review of International Geographical Education Online*, 5, 199–217.

Lane, R. & Caldis, S. (2018) Participatory action research: A tool for promoting effective assessment and building the pedagogical content knowledge of secondary geography teachers, *Geographical Education*, 31, 16–30.

National Committee for Geographical Sciences [NCGS] (2018) *Geography: Shaping Australia's future*, Canberra: Australian Academy of Science.

Shreeve, J. (2018) Addressing the shortage of specialist geography teachers, *Teaching Geography*, 43(3), 98–100.

Timperley, H., Ell, F., Le Fevre, D. & Twyford, K. (2020) *Leading professional learning*, Canberra: ACER.

INDEX

Aboriginal peoples, *see* Indigenous
    Australians
absolute location, 11
absolute space, 10
accessibility (locational concept), 11,
    47–8
accreditation, 294–5
    see also Australian Professional Standards
    for Teachers (APST)
achievement standards (curriculum), 35,
    273, 279
activities, lesson, 133, 139–40
aesthetic values, 25, 44–5, 53–5, 261
affective domain, 145
Africa, 88
Alice Springs (*Mparntwe*) Education
    Declaration (2019), 250
analogue skills (fieldwork), 153–6
analysis
    geographical methods of, 29
    of sources, 280
apps
    difficulties related to, 159
    fieldwork use of, 153, 156, 159, 228–9
    geospatial, 178
    limitations of, 159
Asia region, 44, 59, 61, 66, 74, 89
assessment, 244–7, 273–81
associations, professional, 290, 292–3
atmospheric hazards, 45–6
attribute data, 166–7
Australia
    migration and, 62, 263
    overseas aid from, 91
    population patterns of, 61
    urban future of, 63
    wellbeing variations in, 89–90
Australian Academy of Science, 28
Australian Bureau of Statistics (ABS), 130,
    139–40, 296–7
Australian Capital Territory (ACT), 94

Australian Curriculum
    educational goals informing, 249–50
    general capabilities in, 108–9, 192, 251–4,
    256–63
    HASS learning area in, 249
    History, 4
    multi-dimensional nature of, 250
    Science, 4
Australian Curriculum: Geography
    about, 271
    aims/objectives, 4–5, 131
    assessment and, 273
    choropleth maps in, 121
    concepts in, 6–27
    data skills in, 107, 131
    deconstructing, 271–3
    fieldwork in, 233
    Foundation to Year 6, 36–7
    general capabilities and, 256–63
    graphicacy skills in, 108–12, 119, 121, 124
    on images, 115
    Inquiry and Skills strand, *see* Inquiry and
    Skills strand (curriculum)
    inquiry process and, 187–9
    international curricula and, 290
    key elements, 35
    Knowledge and Understanding strand, *see*
    Knowledge and Understanding strand
    (curriculum)
    population profiles in, 119
    rationale, 131, 204
    senior geography in, 27, 91–6
    spatial technologies in, 164–5
    sustainability defined in, 26
    units in, *see* units
    see also concepts (geographical thinking)
Australian Geography Teachers Association
    (AGTA), 292–3
Australian Professional Standards for Teachers
    (APST), 202, 206, 285–6, 291, 293–5
authorities, fieldwork notification to, 232–3

backward design, 268–9, 282
bar graphs, 118–19
base maps, 149, 155
basic skills (fieldwork), 143–5
belonging, place and, 9
Belt and Road Initiative (China), 74
biodegradable wastes, 24
biodiversity, 77, 97
biology, 4
biomes, 65–7
Biomes and Food Security unit (Year 9), 17, 27, 64–9, 119
biophysical environment, 14–17, 23–5
block diagrams, 59
Bloom's taxonomy, 146–7
budgeting (fieldwork), 235
Bureau of Meteorology (BoM), 41

capabilities
    general (curriculum), 108–9, 192, 251–4, 256–63
    geography and career, 297
career pathways, geography and, 297–8
cartograms, 73, 87
cartographic scale, 21
case studies
    as assessment task, 281
    Biomes and Food Security unit (Year 9), 65–9
    Changing Nations unit (Year 8), 60–3, 263
    Environmental Change and Management unit (Year 10), 77–80
    Geographies of Human Wellbeing unit (Year 10), 85–91, 263
    Geographies of Interconnections unit (Year 9), 71–5
    Global Transformations unit (Year 12), 98
    Land Cover Transformations unit (Year 12), 97
    Landforms and Landscapes unit (Year 8), 53–9, 261
    Natural and Ecological Hazards unit (Year 11), 92
    Place and Liveability unit (Year 7), 46–51
    Sustainable Places unit (Year 11), 96
    Water in the World unit (Year 7), 39–46
    Year 7, 39–52
    Year 8, 53–63, 261, 263
    Year 9, 65–75
    Year 10, 77–91, 263
    Year 11, 92, 96
    Year 12, 97–8
'caterpillar' method of walking, 241
causal relationships, 12, 18, 21, 29
census data, 139–40
central business districts (CBDs), 11
centrality (locational concept), 11
change (concept)
    about, 22
    applications, 22
    characteristics, 7
    spatial distributions and, 12, 22
    units utilising, 23, 53–69
Changing Nations unit (Year 8)
    about, 59–63
    concepts in, 14, 59
    ethical understanding in, 263
    graphs in, 119
    maps in, 121
    urbanisation and development in, 22
checklists, fieldwork, 240
China, 61, 74
cholera outbreak, 167
choropleth maps, 120–1
cities, 22, 63, 81, 83–4, 90
    see also urbanisation
City Road podcast, 290–1
classroom activities, 133, 139–40
climate change, 68, 78, 97
coastal environments, 81–2, 84
cognitive skills/processes, 145–7, 170, 194–5
    see also higher order thinking
collaboration, student, 150, 223, 278–9
communication, 106, 241
communication systems, 11, 72–3

community identity, liveability and, 50
concepts (geographical thinking)
  about, 6–27
  characteristics, 6–7
  as curricular aim, 5
  importance, 27–30
  powerful knowledge and, 30, 207
  spatial technologies and, 171
  see also specific concepts
conservation, 83–4
constructivism, 191–2
  see also inquiry-based learning
content
  knowledge of, 170–1, 173, 206
  pedagogy and, 201–2
  see also Knowledge and Understanding
    strand (curriculum)
cooperative learning, 150, 223, 278–9
Country (Aboriginal), 9, 54–5
COVID-19, 262, 291
critical and creative thinking general
  capability, 253, 259, 261–2
critical thinking, 30
  see also higher order thinking
crop yields, 67
cross-curriculum priorities, 44–5
cultural values, 44–5, 53–5, 261
curiosity, 5, 194–5
curriculum
  deconstructing, 271–3
  knowledge of, 206
  spiral, 38
  see also Australian Curriculum
curriculum-making, 210, 269–73
  see also planning, unit
cyclones, 46

data
  analysing, 166–7, 181
  attribute, 166–7
  classifications, 130, 132
  cleaning the, 132
  in curriculum, 130–3

  defined, 130
  formatting, 180
  lessons related to, 133, 139–40
  literacy around, 129
  mapping, 167
  organising, 28
  recording, 131, 149–50
  representing, 113–22, 132–8, 181
  scale and, 138–40
  sourcing, 179–80
  spatial, 180–1
  terms associated with, 130
  types, 130–1
data collection
  approaches to, 131–2
  digital, 159
  note-taking for, 149–50
  observation as, 148
  sampling methods, 150–1
data visualisations, 134–5, 167, 170, 280
  see also geographic information systems
    (GIS)
DEA Hotspot map (Geoscience Australia),
  174–5
decoding graphics
  as assessment task, 280
  Australian Curriculum on, 111
  cognitive processes used in, 107
  defined, 107
  in fieldwork, 154
  for population profiles, 119
  strategies for, 116, 125
deconstruction, curriculum, 271–3
degradation, environmental, 56, 78
departments, education, 231
deposition, 81
desertification, 56
development, 22, 84, 88–9, 134
diagrams, 59, 122
  see also decoding graphics; encoding
    graphics
Different View, A (2009 GA manifesto), 4
differentiation, 278

digital skills (fieldwork), 153, 156–7
digital technologies, 153, 156–7, 159–60, 228–9
    see also apps; spatial technologies
        (classroom use of); spatial technologies
        (general)
direct instruction, 192–3
direct skills (fieldwork), 143
direction, identifying, 154
distance, effects of, 11
drones, 159
drought, 45
dryland salinisation, 19–20

earth.nullschool website, 134
earthquakes, 58–9
economic values, 44–5
ecosystem services function, 25
ecosystems, 65–6, 77
education, place and, 9
educational objectives, taxonomy of, 108
elaborations (content descriptor), 35
emergent properties (reflexivity), 211
emotional values, 25
employment, 9, 62
encoding graphics
    as assessment task, 280
    Australian Curriculum on, 112
    cognitive processes used in, 107
    population profiles, 119
    strategies for, 122–4
engineering, coastal maintenance, 82
environment(s) (concept)
    changes to, 19, 77–8, 81–2
    characteristics, 6–7
    definition, 14
    functions of, 15, 23–5
    management, 78–80, 82–4
    natural hazards in, 17
    people and, 15–17, 19
    protection of, 57
    quality, 48
    resources from, 24, 39–46
    types of 80

understanding, 16
units utilising, 17, 38–46, 53–9, 64–9,
    76–80
    see also specific environments
Environmental Change and Management
    unit (Year 10), 17, 27, 76–80
environmental determinism, 16
environmental sustainability, 23
equipment, fieldwork, 239–40
erosion, 81
ethical understanding general capability,
    254, 262–3
excursions, 220
experiential learning, 220
explanations, geographical, 29
explicit instruction, 192–3

facilities, access to, 8, 51
feedback, 276–7
    see also assessment
field sketches, 154
fieldwork
    administrative requirements for, 230–3,
        240, 242
    alternatives to, 246–7
    assessments (products), 99, 244–7, 281
    benefits, 220–2
    budgeting, 235
    centrality of, in teaching, 219–20, 223
    checklists, 240
    communication during, 241
    core skills in, 152–7
    culture of, 220
    curriculum placement of, 233
    data collection for, 131, 148–51, 159
    designing tasks for, 145, 223–6
    distance/duration progression in, 224–5
    equipment for, 239–40
    first-aid considerations, 239, 241–2
    geographers and, 218–19
    geographical thinking facilitated by, 147–51
    inhibitors, 229–30
    learning through, 143–7, 150, 219–24

non-attendance, 246
paperwork for, 230–3, 240
pedagogy of, 221–9
planning for, 221, 224, 229–47
in practice, 240–3
pre-fieldwork reconnoitre, 234–5
preparation, 236–40
prior learning for, 235
in school context, 218–19
secondary sources relevant to, 233–4
sequentiality of skills in, 147
staff considerations, 238–9, 241, 243
supervision, 243
teaching, strategies for, 226
technologies supporting, 153, 156–7,
    159–60, 228–9
tools, 154, 157–60
transferable skills from, 143–7
in units, 40–1, 48
see also spatial technologies (classroom
    use of)
fieldwork booklets, 149, 244–6
fieldwork reports, 99, 244–5
first-aid considerations (fieldwork), 239, 241–2
fishing, conservation and, 84
flow maps, 73
flows (interconnection), 19
food production, 66–9
food security, unit on, *see* Biomes and Food
    Security unit (Year 9)
forms (fieldwork planning), 231–3, 240
frameworks, learning (spatial technologies),
    172–5

GapMinder website/app, 87, 134, 296
gender, wellbeing and, 87
general capabilities (curriculum), 108–9, 192,
    251–4, 256–63
generalisations, geographical, 29–30
geographers, fieldwork and, 218–19
    see also fieldwork
geographic information systems (GIS), 131,
    157, 159, 166–7, 179

Geographical Association (GA) (UK), 4, 290
geographical knowledge, *see* Knowledge
    and Understanding strand
    (curriculum)
Geographies of Human Wellbeing unit
    (Year 10)
    about, 85–91
    concepts in, 14, 22
    ethical understanding in, 263
    population profiles in, 119
Geographies of Interconnections unit
    (Year 9), 21, 70–5, 261–2
geography
    career pathways and, 297–8
    coherence of, 28
    in curriculum, *see* Australian Curriculum:
        Geography
    data in, *see* data
    defining, 3–4, 202, 204–5
    as a discipline, 205, 289
    distinctiveness of, 27–8
    fieldwork in, *see* fieldwork
    grammar of, 4
    graphic literacy in, *see* graphicacy
    inquiry-based learning in, *see* inquiry-
        based learning; Inquiry and Skills strand
        (curriculum)
    interdisciplinary nature of, 191–2, 289
    origin of word, 4
    personal, 70–2
    physical, 10, 16
    relevance of, 203–5, 287
    as school subject, 289–90
    teachers' relationships with, 202–3
    teaching, *see* pedagogy/-ies; teaching
        geography
GEOGstandards, 202, 204, 208–9, 285,
    288–92
geomorphology, unit on, *see* Landforms and
    Landscapes unit (Year 8)
Geoscience Australia, 174
geospatial, term, 164–5
geospatial data, 130–1

geospatial skills, 176
geospatial technologies, 164, 168–9
 see also geographic information systems
 (GIS); global positioning systems (GPS);
 spatial technologies (classroom use of)
global navigation satellite systems (GNSS),
 130, 156, 159, 165–6
global positioning systems (GPS), 130, 156,
 159, 165–6
Global Transformations unit (Year 12), 21,
 98–9
globalisation, 72–3
Google Earth, 178
Google Maps, 178
grades, 276
graphicacy
 in Australian Curriculum, 108–12, 119,
 121, 124
 defining, 106
 importance of, 105, 108
 teaching, strategies for, 122–5
 types, 111–22
 visual literacy/thinking and, 107, 111–12
graphical excellence, principles, 123–4
graphics, 113–22, 137
 see also decoding graphics; encoding
 graphics
graphs, 113–14, 117–19
gross national income (GNI) map, 86
guidelines (fieldwork), 231

hands-on learning, 220
hard engineering, 82
hazards, 45–6, 57–9, 92–5
higher order thinking
 about, 30
 fieldwork's development of, 143–7
 graphicacy and, 107, 125
 literacy and, 106
holistic thinking, interconnections and, 19
Hong Kong (China), 61
housing density, 51, 63
human–environment system framework, 19
human wellbeing, see wellbeing

Humanities and Social Sciences (HASS)
 curriculum, 110
Humanities and Social Sciences (HASS)
 learning area, 249
hydrological hazards, 45–6

identity, place and, 9, 50
images, 113, 115–16, 154
 see also decoding graphics; encoding
 graphics
immigration, see migration
India, 89
Indigenous Australians
 Country/Place attachment of, 9, 54–5
 environmental management by, 80
 movements of, 62
 on Uluru, 261
 water resources used by, 45
 wellbeing of, 90
indirect skills (fieldwork), 144
Indonesia, 59–60
Industrial Revolution, 78
infographics, 135–8, 280
information, understanding graphic, see
 graphicacy
Information and communication technology
 (ICT) general capability, 253, 258
inland water environments, 81–2, 84
inquiry-based learning
 Australian Curriculum and, 187–9
 constructivism and, 191–2
 creating a need to know in, 194–5
 data collection using, 132
 direct instruction and, 192–3
 fieldwork and, 219
 flexibility within, 188
 geographical learning and, 190–2
 spatial technologies and, 171
 stages of, 188, 196
 thinking skills developed in, 194–5
 see also fieldwork
inquiry questions
 Foundation to Year 6, 36–7
 purpose, 35, 188

using, 196–7
Year 7, 38
Year 8, 52
Year 9, 64
Year 10, 76
Inquiry and Skills strand (curriculum)
    Biomes and Food Security unit (Year 9),
       65–9
    Changing Nations unit (Year 8), 60–3
    critical and creative thinking in, 259
    data skills in, 131
    Environmental Change and Management
       unit (Year 10), 77–80
    ethical understanding in, 262
    Geographies of Human Wellbeing unit
       (Year 10), 85–91
    Geographies of Interconnections unit
       (Year 9), 71–5
    Global Transformations unit (Year 12), 98
    graphicacy in, 109
    ICT skills in, 258
    Land Cover Transformations unit
       (Year 12), 97
    Landforms and Landscapes unit (Year 8),
       53–9
    literacy skills in, 257
    Natural and Ecological Hazards unit
       (Year 11), 92
    numeracy skills in, 258
    personal and social capabilities in, 260
    Place and Liveability unit (Year 7),
       46–51
    questions in, *see* inquiry questions
    sequence of inquiry in, 188
    Sustainable Places unit (Year 11), 95
    Water in the World unit (Year 7),
       39–46
    Year 7, 39–52
    Year 8, 53–63, 260
    Year 9, 65–75, 257–8
    Year 10, 77–91, 258–9, 262
    Year 11, 92, 95
    Year 12, 97–8
    see also graphicacy

Institute of Australian Geographers (IAG),
    290
interconnection (concept)
    about, 17
    causal relationships from, 18
    characteristics, 6–7
    concepts related to, 18–19
    generalisations using, 29
    global factors facilitating, 72–5, 261–2
    holistic thinking and, 19
    interdependence and, 18
    units using, 21, 70–5, 261–2
intercultural understanding general
    capability, 254, 260–2
interdependence, 18
internet, 72, 159, 177–8
interviews, conducting, 154
irrigation systems, 44–5, 66
Italy, 75

judgements, quality (assessment), 278

knowledge
    content, 170–1, 173, 206
    curricular aims for, 5, 36–7
    curriculum, 206
    dimensions of, 145–7
    inquiry linked to, 188
    pedagogical content, 205–6, 208–9
    powerful, *see* powerful knowledge
    in TPACK framework, 173–5
    see also Knowledge and Understanding
       strand (curriculum)
Knowledge and Understanding strand
    (curriculum)
    Biomes and Food Security unit (Year 9),
       65–9
    Changing Nations unit (Year 8),
       59–63
    Environmental Change and Management
       unit (Year 10), 77–80
    ethical understanding in, 262
    Geographies of Human Wellbeing unit
       (Year 10), 85–91

Knowledge and Understanding strand
  (curriculum) (*cont.*)
  Geographies of Interconnections unit
    (Year 9), 70–5
  Global Transformations unit (Year 12), 98
  intercultural understanding in, 260–1
  Land Cover Transformations unit
    (Year 12), 97
  Landforms and Landscapes unit (Year 8),
    53–9, 272
  Natural and Ecological Hazards unit
    (Year 11), 92
  personal and social capabilities in, 260
  Place and Liveability unit (Year 7), 46–51
  Sustainable Places unit (Year 11), 95
  Water in the World unit (Year 7), 39–46
  Year 7, 39–52
  Year 8, 53–63, 260, 272
  Year 9, 65–75
  Year 10, 77–91, 260–2
  Year 11, 92, 95
  Year 12, 97–8

land clearance, 20, 81–2
Land Cover Transformations unit (Year 12),
  17, 97
land degradation, 56, 68
land environments, 81–3
landforms
  changes in, 22, 57
  geomorphic processes and, 55–6
  in landscapes, 53
  values attached to, 53–5, 261
Landforms and Landscapes unit (Year 8)
  about, 52–9
  assessment and, 279
  concepts in, 17, 53
  cultural understanding and, 261
  deconstructing, 272–3
landscapes
  geomorphic processes and, 55–6
  human impact on, 56–7
  landform features of, 53

protecting, 57
  unit on, *see* Landforms and Landscapes
    unit (Year 8)
  values attached to, 54–5
learning
  activities, 133, 139–40
  assessing, *see* assessment
  cooperative, 150, 223, 278–9
  fieldwork-related, 143–7, 150, 219–24
  frameworks, 172–5
  inquiry-based, *see* inquiry-based learning
  pedagogical approaches to, *see*
    pedagogy/-ies
  professional, *see* professional learning
  requirements, identifying, 271–3
  student-centred, 171, 223–4
lesson activities, 133, 139–40
lessons, distinctive and powerful
  curriculum-making for, 210
  enacting, 210–15
  protocols and, 212
  reflexivity and, 210–12
  'what?' and 'why?' of, 202–9
  see also planning, unit
letters of explanation (fieldwork), 232
life expectancy, 21, 86
lineal graphs, 113
literacy
  data, 129
  graphic, *see* graphicacy
  higher order thinking and, 106
  visual, 107, 109, 111
literacy general capability, 108–9, 252, 256–7
liveability, 46–51
  see also Place and Liveability unit (Year 7)
loading checklist (fieldwork), 240
location(s)
  absolute, 11
  concepts, 11–12
  effects of, 11
  of geomorphological patterns, 58
  maps for, 149
  relative, 11

map comparison method, 12
map scale, 21
maps
  base, 149, 155
  choropleth, 120–1
  creating, 280
  flow, 73
  geographical use of, 119–21
  geomorphological hazard, 58
  location, 149
  mental, 71
  rainfall distribution, 13
  reading and interpreting, 154, 280
  sketch, 156
  types, 120
  weather, 41
  wellbeing measurements and, 85–7
  see also decoding graphics; encoding
    graphics
marine environments, 81, 83–4
marks (assessment), 276
matrices, fieldwork, 149
Melbourne Declaration on Educational Goals
  for Young Australians (2008), 249–50
mental maps, 71
migration, 60–3, 263
mobile technologies (fieldwork), 228–9
  see also apps
motivation, student, 171
Murray-Darling Basin, 44

NASA WorldWind, 179
national parks, 83
Natural and Ecological Hazards unit
  (Year 11), 17, 92–5
'need to know', creating a, 194–5
New South Wales (NSW), 93
non-biodegradable wastes, 24
non-government organisations (NGOs), 91
non-renewable resources, 24
Northern Territory (NT), 94
note-taking (fieldwork), 131, 149–50
numeracy general capability, 252, 257–8

objects, as inquiry stimuli, 195
observational scale, 21
observations (fieldwork), 148, 155
observations (student assessment), 277
ocean environments, 81, 83–4
online geospatial tools, 177–8
online professional networks, 293
opportunistic sampling method, 151
organisations, wellbeing improvement role
  of, 90–1
out-of-field teachers, 289

Pacific Islands, 88
paperwork (fieldwork), 230–3, 240
parent letters (fieldwork), 232
PCK framework (Shulman), 173
pedagogical content knowledge (PCK),
  205–6, 208–9
pedagogical knowledge (TPACK
  framework), 173
pedagogy/-ies
  constructivist, 191–2
  content and, 201–2
  fieldwork, 221–9
  GEOGstandards of, 202, 204, 208–9
  graphicacy-related, 124
  powerful, 207–9
  spatial technologies, 171–5
peer-assessment (student), 276
people, environment and, 15–17, 19
perceived space, 11
permission forms (fieldwork), 231
personal geographies, 70–2
personal and social general capability, 253,
  259–60
photo logs, 149
photographs, taking, 155
physical geography, 10, 16
place (concept)
  about, 7
  causal relationships and, 18
  characteristics, 6–8, 47
  data and, 28

place (concept) (*cont.*)
    effects of, 8–9
    making and changing of, 7–8
    perceptions of, 47, 70–2
    sustainability and, 27
    units utilising, 10, 27, 46–51, 53–9, 70–2
Place (Torres Strait Islanders), 9
Place and Liveability unit (Year 7)
    about, 46–51
    concepts in, 10, 46
    podcasts relevant to, 291
    population profiles in, 119
planning, fieldwork, 221, 225, 229–47
planning, unit
    approach to, 267–73
    assessment and, 273–81
    backward design/UBD in, 268–9, 282
    deconstruction process in, 271–3
    importance of quality, 267–8
    models for, 269–70
    preparing for, 269–70
    questions guiding effective, 282
podcasts, 290–1
policies
    fieldwork, 231
    geography defined in, 204
pollution, 48, 78, 81
population growth, 68–9, 88, 263
population patterns, comparing, 60–1
population profiles/pyramids, 118–19
powerful knowledge, 30, 205–9, 270
powerful pedagogy, 207–9
PQE strategy, 121
primary data, 130
processes
    fieldwork, 230–1
    interconnection, 18
professional associations, 290
professional engagement
    activities promoting, 287–8
    APST standards and, 285–7, 293–5
    defined, 285–6
    across geographic discipline, 290

geographical tools for, 296–7
with geography beyond the classroom,
    297–8
GEOGstandards on, 285, 289–92
importance of, 287, 298
networks for, 292–3
professional learning and, *see* professional
    learning
purpose, 286
professional learning
    about, 293
    communities, 294
    GEOGstandards on, 292
    opportunities, 285, 293–5
professional standards, *see* Australian
    Professional Standards for Teachers
    (APST); GEOGstandards
Professional Standards for the Accomplished
    Teaching of School Geography, *see*
    GEOGstandards
pro-formas, 149
protocols, lesson development, 212
proximity (locational concept), 12
public transport, 47–8

QGIS (Quantum GIS), 179
quadrats, 149, 155
qualitative data, 130
quantitative data, 130
Quantum GIS (QGIS), 179
Queensland, 93
questions, 28, 149, 280
    see also inquiry questions
quizzes, 280

rainfall, 13, 42, 56
random sampling method, 151
record-keeping (fieldwork administration), 242
record sheets, 149–50, 155
    see also fieldwork booklets
recreation, conservation and, 84
reflexivity, 210–12
relative location, 11

relative space, 10
remote sensing, 168–9
remoteness (locational concept), 12
renewable resources, 24
research inquiry tasks, 132
    see also inquiry-based learning
reserves, 83
resource(s)
    environmental, 24, 39–46
    fieldwork (administrative), 231
    water as, 39–46
respect (curricular aim), 5
rice growing, 44, 66
rivers, 40–1, 81–2
rubrics, 276
runoff patterns, 56
rural areas, 90

salinisation, 19–20
salinity (soil), 68
sampling (fieldwork), 150–1
SAMR (substitution, augmentation,
    modification, redefinition) model, 173–5
scaffolding, 279
scale (concept), 7, 21–2, 138–40
scale (map/cartographic), 21
science, technology, engineering and
    mathematics (STEM), 191–2
secondary data/sources, 130, 132, 233–4, 256
self-assessment (student), 276
senior geography
    Australian Curriculum on, 27, 91–6
    fieldwork reports in, 99
    geospatial skills in, 176
    state curriculum for, 91–4
services, access to, 8
sink function (environment), 24–5
site analysis, 155
sketch maps, 149, 156
skills
    analogue, 153–6
    Australian Curriculum on, *see* Inquiry and
        Skills strand (curriculum)

basic, 143–5
cognitive, 145–7, 170, 194–5
as curricular aim, 5
data-related, *see* data
digital, 153, 156–7
direct/indirect (fieldwork), 143
fieldwork's core, 152–7
geospatial, 176
graphic literacy, *see* graphicacy
higher order, 143–5
sequential development of, 147
transferable, 143–7, 297
SMART goals, 210
smartphone apps, *see* apps
Snow, John, 167
Social Atlases of Australia, 50
social connection, liveability and, 50
societies, geographical, 293
society, spatial distributions and, 14
soft engineering, 82
soil degradation, 68
source function (environment), 24
sources
    analysing, 280
    secondary, 130, 132, 233–4, 256
South America, 88
South Australia (SA), 94
space (concept)
    absolute, 10
    characteristics, 6–7
    conceptualisations of, 10–11
    data and, 28
    locational concepts and, 11–12
    organisation of, 14
    perceived, 11
    relative, 10
    society and, 14
    spatial changes and, 12
    spatial distributions and, 12–14
    units utilising, 14, 39–46, 85–91
spatial changes, 12
spatial distributions, 12–14, 22, 85–90
    see also urbanisation

spatial technologies (classroom use of)
  Australian Curriculum on, 164
  benefits, 170–1
  difficulties related to, 159
  in fieldwork, 159–60
  incorporation and implementation of, 163, 171–81
  limitations, 159
  tools and resources for, 176–81
  see also geographic information systems (GIS); global positioning systems (GPS)
spatial technologies (general), 164, 168–9
  see also geographic information systems (GIS); global positioning systems (GPS)
spatial thinking, 170
speculation, as inquiry stimuli, 195
spiral curriculum (Bruner), 38
spiritual values, 15, 25, 44–5, 53–5, 261
spreadsheets, 180
staffing, fieldwork, 238–9, 241, 243
standards, professional, *see* Australian Professional Standards for Teachers (APST); GEOGstandards
states
  assessment and, 246, 275
  senior geography curriculum in, 91–4
  on spatial technologies, 164
statistics, 85–8, 296–7
Statsilk (statistical tools), 296
STEM (science, technology, engineering and mathematics), 191–2
stratified sampling method, 151
student-centred learning, 171, 223–4
students
  fieldwork benefits for, 221–2
  information about (fieldwork), 231–3, 240
  motivation, 171
substitution, augmentation, modification, redefinition (SAMR) model, 173–5
surveys, 48, 156
sustainability (concept)
  applications, 26
  characteristics, 7
  in cities, 63, 84

definitions, 23, 26, 77–8
disagreements about, 25–6
environmental functions and, 24–5
evaluation using, 30
place and, 27
units utilising, 27, 64–9, 76–80, 85
world-views on, 78–9
Sustainable Places unit (Year 11), 10, 27, 95–6
system function (environment), 25
systematic sampling method, 151
systems
  environmental, 81–2
  interconnection, 19–20

tables, fieldwork, 149
Tasmania, 94
taxonomy of educational objectives, 108
taxonomy of graphs, 113–14
teachers
  accreditation, 294–5
  assessment suggestions for, 275–6
  fieldwork roles of, 223, 226
  geographic subject identity of, 202–3
  as inquirers, 194
  knowledge development in, 206
  out-of-field, 289
  planning and design role of, 267–9
  reflexivity and, 210–12
  standards for, *see* Australian Professional Standards for Teachers (APST); GEOGstandards
  see also fieldwork
teaching geography
  fieldwork benefits for, 221–2
  fieldwork teaching, strategies, 226
  frameworks for, 172–5, 205–6
  graphicacy strategies, 122–5
  inquiry in, *see* inquiry-based learning; Inquiry and Skills strand (curriculum)
  lessons for, *see* lessons, distinctive and powerful
  methods, subjectivity around, 35
  pedagogical approaches, *see* pedagogy/-ies
  powerful knowledge in, 205–7

spatial technologies, strategies for, 172–81
standards for, *see* GEOGstandards
technological knowledge (TPACK framework), 172
technological, pedagogical and content knowledge (TPACK) framework, 172–5
technology/-ies
difficulties related to, 159
in fieldwork, 153, 156–7, 159–60, 228–9
limitations of, 159
spatial, *see* spatial technologies (classroom use of); spatial technologies (general)
territories
assessment and, 246, 275
senior geography curriculum in, 91–4
on spatial technologies, 164
texts, geographic, 130, 132, 233–4, 256
thinking, geographic, *see* concepts (geographical thinking)
thinking skills/processes, 145–7, 170, 194–5
see also higher order thinking
time-space convergence, 11
tools
fieldwork, 154, 157–60
geographical, 296–7
geospatial, 176–81
topophilia, 142
Torres Strait Islander peoples, *see* Indigenous Australians
tourism, 74–5, 84
TPACK (technological, pedagogical and content knowledge) framework, 172–5
trade, 73–4
transects, 149, 152, 156
transportation systems, 11, 72
trip intention forms (fieldwork), 232–3
tropophilia, 142
tsunamis, 58
Tufte's principles of graphical excellence, 123–4

Uluru, climbing of, 261
understanding, geographical, 188, 190
see also Knowledge and Understanding strand (curriculum)

understanding by design (UBD), 268–9, 282
see also planning, unit
United States, 61
units
assessment and, 279–81
concepts, linking of, 28
deconstructing, example, 272–3
planning for, *see* planning, unit
Year 7, 38–52
Year 8, 52–63
Year 9, 64–75
Year 10, 76–91
Year 11, 92–5
see also specific units
universities, partnerships with, 290
urban areas, 22, 63, 81, 83–4, 90
urban sprawl, 63
urbanisation
causes and consequences of, 60
development and, 22
management and planning of, 63
migration and, 61–3
patterns related to, 60–1
units exploring, 59–63

value(s)
environmental functions and, 25
of landscapes and landforms, 53–5, 261
resource use and, 44–5
vegetation, 20–1, 81–2
Venice (Italy), 75
Victoria, 93
videos, making, 156
visual literacy/thinking, 107–8, 111
visualisation skills, 170
visualisations, data, *see* data visualisations
volcanoes, 58

wastes, 24–5
water, 43–4, 56, 82, 84
water features, 40–1
Water in the World unit (Year 7), 17, 38–46, 121

weather, 41–2, 134
websites, data visualisation, 73, 87, 134–5, 296
wellbeing
   in Australia, 89–90
   development and, 88–9
   improving, 91
   in India, 89
   measuring and mapping, 85–8
   place and, 9
   scale and, 22

   unit on, *see* Geographies of Human
     Wellbeing unit (Year 10)
Western Australia (WA), 93
windows
   in geography pitch development, 203
   in inquiry process, 195, 197
wonder (curricular aim), 5
worksheets, 149–50, 155
   see also fieldwork booklets
WorldMapper website/app, 73, 87